Nature-Based Solutions for Urban Renewal in Post-Industrial Cities

EDITED BY SILVIA BARBERO AND AXEL TIMPE

NEW YORK AND LONDON

Designed cover image: © Sabina Reichert

First published 2025
by Routledge
605 Third Avenue, New York, NY 10158

and by Routledge
4 Park Square, Milton Park, Abingdon, Oxon, OX14 4RN

Routledge is an imprint of the Taylor & Francis Group, an informa business

© 2025 selection and editorial matter, Silvia Barbero and Axel Timpe; individual chapters, the contributors

The right of Silvia Barbero and Axel Timpe to be identified as the authors of the editorial material, and of the authors for their individual chapters, has been asserted in accordance with sections 77 and 78 of the Copyright, Designs and Patents Act 1988.

The editors acknowledge the contribution of Alessandro Campanella and Cecilia Padula as Co-Editors of this volume.

The Open Access version of this book, available at www.taylorfrancis.com, has been made available under a Creative Commons Attribution (CC-BY) 4.0 license.

Any third party material in this book is not included in the OA Creative Commons license, unless indicated otherwise in a credit line to the material. Please direct any permissions enquiries to the original rightsholder.

Trademark notice: Product or corporate names may be trademarks or registered trademarks, and are used only for identification and explanation without intent to infringe.

ISBN: 9781032756233 (hbk)
ISBN: 9781032756219 (pbk)
ISBN: 9781003474869 (ebk)

DOI: 10.4324/9781003474869

Typeset in Dante and Avenir
by codeMantra

Nature-Based Solutions for Urban Renewal in Post-Industrial Cities

This book, based on the experiences and insights gained during the Horizon 2020 project proGIreg, offers a detailed overview of targeted nature-based solutions and their impacts on various key sustainability areas, guiding readers through the spatial analysis, co-design, and implementation processes of cities in Europe and Asia. Chapters shed light on the challenges and opportunities encountered in each location, including Germany, Italy, Croatia, Bosnia and Herzegovina, Greece, Portugal, Romania, and China. It also shares essential lessons learned and a wide range of indicators crucial for assessing the benefits of nature-based solutions on social innovation, circular economy, biodiversity, and health. Finally, the focus of this book shifts to the future of nature-based solutions as catalysts for new and green community economies as well as policies aimed at addressing climate change and urban renewal. The lessons and insights from the projects highlighted in this book will be valuable for urban planners and policymakers worldwide, as well as for a broader audience interested in nature-based solutions and urban regeneration.

Silvia Barbero is an Associate Professor at the Department of Architecture and Design of Politecnico di Torino (Italy). She is an expert in systemic design, focusing on industrial innovation and territorial enhancement to promote environmental, social, and economic sustainability. Additionally, she actively contributes to academic research and education.

Axel Timpe is a landscape architect and Associate Professor at the Institute of Landscape Architecture at RWTH Aachen University. His research and teaching is focused on green infrastructure, urban agriculture, and nature-based solutions and their co-production with local communities. He has been the principal investigator of the proGIreg Innovation Action.

Contents

Acknowledgements	vii

1 Introduction: Nature for renewal — 1
Axel Timpe and Silvia Barbero

2 NBS for social innovation — 8
Mais Jafari, Rolf Morgenstern and Jonas Runte

3 NBS for circular economy: baselines for sustainable NBS implementation — 24
Axel Störzner, Kimberly Schnell and Theresa Kemeny

4 NBS for biodiversity — 36
Simona Bonelli, Federica Larcher, Lingwen Lu, Marta Depetris and Manuela Ronci

5 Dortmund Living Lab — 57
Margot Olbertz and Mais Jafari

6 Zagreb Living Lab — 78
Bojan Baletić, Iva Bedenko and Marijo Spajić

7 NBS spatial analysis processes: the role of spatial analysis for Front-Runner Cities and Follower Cities — 101
Sabina Reichert, Oana Emilia Budău and Codruț Papina

vi Contents

8 Co-design NBS with post-industrial communities 121
Margot Olbertz, Bettina Wilk, Israa Mahmoud,
Emanuela Saporito and Ina Säumel

9 Turin Living Lab 152
Silvia Barbero and Federica Larcher

10 Ningbo Living Lab 165
Ruowen Wu, Tian Ruan and Yaoyang Xu

11 Evidence-based NBS benefits and related indicators 179
Chiara Baldacchini, Carlo Calfapietra and Martina Ristorini

12 Benefits from social innovation 189
Egidio Dansero, Luca Battisti, Federico Cuomo,
Giacomo Pettenati, Giovanni Sanesi and Giuseppina Spano

13 Benefits from circular economy 207
Rolf Morgenstern and Bernd Pölling

14 Benefits for biodiversity 215
Simona Bonelli, Lingwen Lu, Marta Depetris, Tian Ruan,
Monica Vercelli, Ruowen Wu and Yaoyang Xu

15 Benefits for health and wellbeing 230
Mònica Ubalde López, and Payam Dadvand

16 Unlocking potential: Follower cities' NBS replication
strategies for greening urban environments 247
Margot Olbertz, Codrut Papina, Athina Abatzidi,
Melania Blidar, Sandra Dimancescu, Helga Gonçalves,
Violeta Irimies, Vasiliki Manaridou, Amra Mehmedić,
Teresa Ribeiro, Mirza Sikirić, Bogdan Stanciu and Nerantzia Tzortzi

17 NBS policymaking at the forefront: NBS for change
and resilience 285
Karin Zaunberger

18 Business models in NBS 301
Bernd Pölling and Rolf Morgenstern

Index 325

Acknowledgements

This document has been prepared in the framework of the European project proGIreg.

The sole responsibility for the content of this publication lies with the authors. It does not necessarily represent the opinion of the European Union. Neither the REA nor the European Commission is responsible for any use that may be made of the information contained therein.

This project has received funding from the European Union's Horizon 2020 research and innovation programme under grant agreement no. 776528.

Introduction

Nature for renewal

1

Axel Timpe and Silvia Barbero

Nature-based solutions to urban regeneration challenges

With the transition from the industrial to a post-industrial era, human societies face multifaceted challenges. We need to organise the transition from the fossil-based energy system to a new energy regime which relies on renewable energy sources as did the solar energy system in the pre-industrial era. This will also strongly transform bio-based production like agriculture which in its mechanisation and fertilisation today is still fuelled by fossil energy. The post-industrial society will have to deal with the leftovers of the industrial era as well on the global as on the local level. Climate change induced by CO_2 and other greenhouse gas emissions will impact societies and ecosystems around the world and require considerable adaption efforts. The material emissions of industries can spread globally, as in the case of microplastics, and also have very concentrated local impacts like pollution and depletion of local natural resources like soils and waterbodies. Pollutants from the industrial economy will continue to impact the human environment and health for centuries. Finally, after an industrial era perceived as a period of wealth and wellbeing by many citizens living in industrialised countries, a post-industrial future will have to prove how to maintain and enhance people's quality of life and secure livelihoods in transforming economies. Overall these challenges are thus appropriately being described as the great transformation.

Cities are in the focus of this transition. Especially in cities with a rich and successful industrial past, the different transformation challenges accumulate. These cities have to deal with the polluting leftovers of industries, derelict land

DOI: 10.4324/9781003474869-1

This chapter has been made available under a CC-BY license.

2 Nature-Based Solutions for Urban Renewal in Post-Industrial Cities

which hasn't found a new vocation yet, and a population which, once attracted by the opportunities of industrial success, now needs to find new opportunities of decent work and life. Having profited from a thriving economy, full employment, and a welfare state at the peak of their industrial evolutions, these cities now have to face multiple challenges at a time: Environmental challenges are, among others, few greenspaces, poor and arid soils, soil pollution, spatial barriers and inaccessibility, and a strong heat island effect intensifying with climate change. Social challenges include few economic opportunities and high unemployment rates of the local population, low educational opportunities leading to an ageing population, and a social segregation between and within different urban areas. While these factors already present considerable risks to human health, post-industrial cities and their urban populations are also most vulnerable to the effects of climate change and thus have a high need for adaption to the hanging climate conditions.

A neglected factor during the industrial era has been nature. Natural capital has provided the resources for most industrial processes and has long been considered as free and inexhaustible. Economically speaking, natural capital and also the negative impacts of industrialisation on nature have been treated as externalities that were not part of the economic equations the success of the industrial society relied on.

In the transformation from the industrial to a post-industrial society, nature again has an important role to play, but this time as an active and highly valued part of the equation. Nature can be considered an important part of the solution for the ongoing and upcoming transformation challenges. For this the term nature-based solutions (NBS) has been coined and shall become an important part of transformation strategies in the European Union (EU) and worldwide.

The European Commission defines NBS as

> Solutions that are inspired and supported by nature, which are cost-effective, simultaneously provide environmental, social and economic benefits and help build resilience. Such solutions bring more, and more diverse, nature and natural features and processes into cities, landscapes and seascapes, through locally adapted, resource-efficient and systemic interventions.
>
> Nature-based solutions must therefore benefit biodiversity and support the delivery of a range of ecosystem services.
>
> (European Commission, 2020)

NBS rely on living ecosystems. These ecosystems can be found and used in a broad variety, and NBS can thus rely on the better use and protection

Introduction: Nature for renewal

of natural or protected ecosystems as well as on managed or restored ecosystems and even on the design and management of newly created ecosystems (Eggermont et al., 2015). In the context of post-industrial cities and their highly transformed environments the focus is of course on restored and newly created ecosystems as NBS.

Living Labs for nature-based solutions

While applied at the large scale of global bioregions or natural landscapes NBS can help mitigating climate change or its effects like desertification and instability of coastlines, this book concentrates on the question how NBS can be used for supporting the post-industrial transformation of cities, even with a more detailed interest in urban regeneration areas at the city district level. It relies on a Living Lab research that has been carried out between 2018 and 2023 in four cities in Europe and China (Figure 1.1), each of them having its specific industrial history.

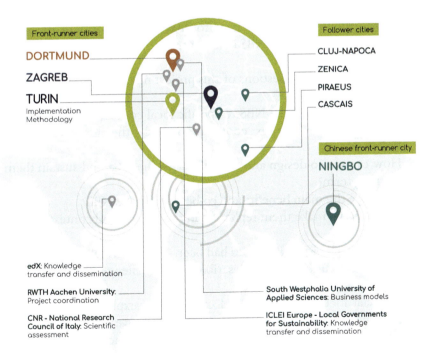

Figure 1.1 The proGIreg research cities. Credit: proGIreg/RWTH Aachen University, Institute of Landscape Architecture.

4 Nature-Based Solutions for Urban Renewal in Post-Industrial Cities

Dortmund in Germany as part of the Ruhr region has a long history of early industrialisation in the 19th century with coal and steel production. Coal mining having disappeared from the region since the 1970s with the last mine closed in 2015, steel production is still present, but of much smaller importance than historically. The local Living Lab is located around a former coal mine closed in 1980 and the associated coking plant closed in 1992.

Turin in Italy has been the capital of the Italian car industry, one of the most important industry sectors in the second half of the 20th century. While the industry is still present in the city, the production and the involved workforce have been strongly reduced, the district around the formerly largest car factory now being the Living Lab for NBS.

Zagreb as the capital of Croatia is encountering post-industrial transformation and growth at the same time. The borough of Sesvete serving as Living Lab has lost its large-scale meat production and processing industry since the end of the Yugoslavian Federation in 1992 and accommodated a quickly growing population partly composed of Croatian refugees of war in Bosnia and Herzegovina, who resettled here.

Ningbo in China is a harbour city which still experiences urban and industrial growth, but has the need to improve the quality of life and historic green spaces through NBS in its centre with the Moon lake and its park serving as a Living Lab.

The main research questions of this project have been:

- How can we adapt the NBS to specific local contexts of the mentioned post-industrial cities in cooperation with the local citizens and stakeholders?
- How can we co-design and co-implement the NBS and sustain them in the long term?
- What are the main benefits and co-benefits of the NBS?
- How can we make them replicable in other cities and contexts?

For this research eight NBS types had been selected which were promising, especially as solutions to the described local challenges of post-industrial transformation (Figure 1.2).

A special focus of the NBS selection lies in exploring the productive ecosystems services of nature. Especially NBS 1–4 have the intention to create a productive output which can help creating new economies and supporting livelihoods based on nature in post-industrial areas. Making landfills as a leftover of the industrial past accessible is combined with the production of renewable energy in NBS 1. NBS 2 uses local resources, excavation

Introduction: Nature for renewal   5

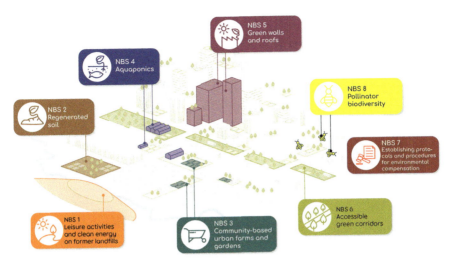

Figure 1.2 Eight types of NBS for post-industrial urban regeneration have been tested in proGIreg. Credit: proGIreg/RWTH Aachen University, Institute of Landscape Architecture.

material from construction sites, and urban compost and combines them with microorganisms to create new, fertile soil for urban greening in a natural process. Local citizens find the opportunity to produce their own food and experience contact with nature as part of urban gardening and farming initiatives in NBS 3. This agricultural production is also the goal of NBS 4, but with a more controlled ecosystem created for the combined production of fish and vegetables in aquaponics. As soil-less agriculture this NBS is also being tested on polluted formerly industrial sites. All these productive NBS rely on local resources and have a circular economy dimension.

Due to this special focus, the research project, which has been supported by the European Union's Horizon 2020 programme as an Innovation Action with 10.5 mio Euro, has been named "productive Green Infrastructure for post-industrial urban regeneration (proGIreg)". The named productive NBS have been complemented by four additional NBS to allow an integrated regeneration of post-industrial areas in cities and add a larger set of main benefits intended. Green walls and green roofs (NBS 5) as well as green corridors (NBS 6) help to adapt cities to climate change and with NBS 8 for pollinators have a special focus on biodiversity. Overall, the of the NBS put to a test in Living Labs cover a wide array of targeted benefits and co-benefits as shown in Figure 1.3.

This book gives deeper insights into the potential of NBS to regenerate post-industrial urban neighbourhoods to make the experience of the

6 Nature-Based Solutions for Urban Renewal in Post-Industrial Cities

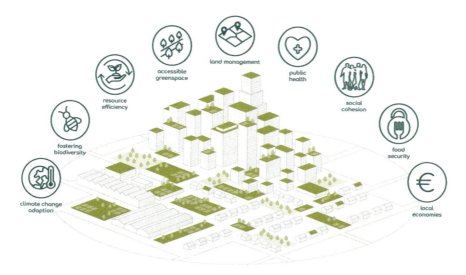

Figure 1.3 NBS can provide a wide array of benefits to post-industrial urban areas. Credit: proGIreg/ICLEI.

proGIreg project accessible to anyone who wants to replicate this in other contexts. It, therefore, presents two types of chapters for presenting local experiences as well as theoretical and scientific findings. The former provide insights into the Living Lab processes and results in the cities already mentioned and, in addition, to the planning processes in four additional cities Cascais (Portugal, Cluj-Napoca (Romania), Piraeus (Greece), and Zenica (Bosnia and Hercegovina) which will replicate the NBS in the next years. The other chapters present the theoretical and scientific insights gained through the research process. These experiences from cities with and without Living Labs are presented in Chapters 5, 6, 9, 10, and 16.

Chapter 2 begins by outlining some of the main areas of NBS intervention in post-industrial cities, described earlier: the creation of a productive and circular economy, the improved biodiversity, and the social innovations necessary when creating NBS. As a second theory block, Chapters 7 and 8 explain how the principle of local adaption of NBS can be met by participatory planning processes from the spatial analysis approach to a co-design process with local communities. NBS are used to solve human or societal problems. The research carried out in proGIreg has created guidelines how to monitor the benefits of NBS in addressing specific challenges of urban regeneration areas and provides an evidence base on important key performance indicators in this field. Chapters 11–15 present these findings.

The research findings lead to the conclusion that NBS can become an important tool for transformation, as well on the strategic level of the EU and its policies as on the local level of municipalities and local communities. Chapters 17 and 18 present both perspectives: the EU policymaking for NBS described by a member of the European Commission and the opportunities for local initiatives to start NBS by using business and governance models tested in the proGIreg Innovation Action.

The editors thank all authors for their e-contributions to this volume for their commitment during the five years research process. Only in a joint effort it was possible to bring NBS into existence in many different places and to reflect on the benefits they provide and on how they can be replicated in other places in Europe and around the world. We hope that through this book other cities and initiatives will be inspired to integrate NBS into the urban regeneration strategies they are setting up for their local context and challenges.

References

Eggermont, H., Balian, E., Azevedo, J. M. N., Beumer, V., Brodin, T., Claudet, J., Fady, B., Grube, M., Keune, H., Lamarque, P., Reuter, K., Smith, M., van Ham, C., Weisser, W. W., & Le Roux, X. (2015). Nature-based Solutions: New Influence for Environmental Management and Research in Europe. *GAIA - Ecological Perspectives for Science and Society, 24*(4), 243–248. https://doi.org/10.14512/gaia.24.4.9

European Commission. (2020). *Nature-based Solutions.* https://ec.europa.eu/info/research-and-innovation/research-area/environment/nature-based-solutions_en

NBS for social innovation **2**

*Mais Jafari, Rolf Morgenstern
and Jonas Runte*

Social innovation and nature-based solutions

In the wake of increasing natural disasters, global warming and the loss of biodiversity, worldwide consensus has emerged that climate change requires calls for counteractive measures to be taken. This is expressed, for example, by the Intergovernmental Panel on Climate Change Report (IPCC), the Paris Climate Agreement, or the Sustainable Development Goals (Brondizio et al., 2019). Yet, at the same time there is a controversial debate on how climate change mitigation should be designed. Nature-based solutions (NBS) approach especially receives a high level of advocacy to limit global warming by 2 degrees (Herlyn, 2021). The European Commission regards NBS as key combining human livelihoods, economic prosperity, and ecological diversity, and need to be promoted (European Commission, 2015). Furthermore, NBS embodies new approaches to socio-ecological resilience and support social innovation through integrating socio-ecological systems associated with academic dialogue into urban projects to address various environmental challenges while providing multiple co-benefits to the economic, social, and ecological domains (European Commission, 2021). However, social innovations are characterized by great complexity, which not only extends over the process of NBS implementation, but especially includes diffusion, where citizens adopt social innovations as a new practice and routine. In Dortmund, five NBS support to realize more intense citizen participation during co-design, co-implementation, and co-management. One of the biggest challenges of co-design is the high

DOI: 10.4324/9781003474869-2

This chapter has been made available under a CC-BY license.

NBS for social innovation **9**

variation in the degree of citizen participation reaching from information to empowerment, depending on the respective NBS layout. Civil society is not equally involved in the co-creation of all NBS, so liability aspects during construction and maintenance require involvement of experts such as the movement park (NBS1). Certain NBS involve technical solutions, such as the aquaponic system (NBS4) and the path connection at Deusenberg (NBS6), which consequently cannot be planned and operated by everyone (Ayob et al. 2016).

Given this, the question arises how NBS become social innovations and how they succeed. An analysis will be carried out that provides information on the transition from an NBS to a social innovation. Moreover, insights will be presented which factors make social innovation successful.

Case studies, NBS in Dortmund Living Lab

To understand the dimensions of social innovation in green infrastructure projects, this chapter analyzes the co-creation process of NBS in the Dortmund Living Lab (LL). Special focus will be given to two NBS: community-based urban farms and gardens (NBS3) and pollinator biodiversity in the Dortmund LL (NBS8). The application of community capacity building, interdisciplinary synergies, communication and interaction between actors as well as digital technologies are strongly incorporated in these NBS.

Dortmund LL is located about 4.5 km northwest of Dortmund city center within Huckarde district (one of Dortmund's 12 districts). Since 1850, with the existence of the coal mine and coking plant the area prospered. The closing of the coal mine in 1980 and of Hansa Coking Plant in 1992 left more than 9,000 people without jobs leading to tremendous economic, social, and environmental problems in Huckarde settlement. Huckarde still faces multiple regeneration challenges in regard to environmental degradation of post-industrial sites, socioeconomic disparities, and partly poor access to Deusenberg recreation area (former landfill east of Huckarde settlement) or to Dortmund downtown due to still-used rail infrastructure. Today, about 9,150 citizens live within the LL.

The goal was to develop a network of green infrastructure through the implementation of five NBS (Figure 2.1) to improve connectivity, thus enhancing living and environmental conditions in Huckarde, while disseminating and replicating these solutions and practices in other locations in Dortmund and other national and international contexts.

10 Nature-Based Solutions for Urban Renewal in Post-Industrial Cities

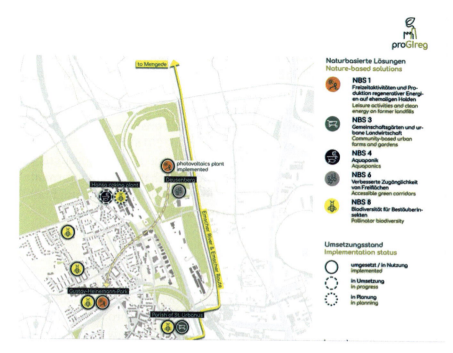

Figure 2.1 Locations of NBS in Dortmund LL.
Source: ProGIreg, November 2022

NBS 3: Community-based urban farms and gardens

Part of the LL was dedicated for the creation of a permaculture food forest. The term permaculture is derived from the term permanent agriculture. It refers to a sustainable agriculture that forms a closed cycle, thus maintaining itself over time. In a food forest the creation of a natural forest area is imitated, using mainly edible plants. Creating a food forest requires careful and extensive planning but requires less maintenance to operate than a conventional garden. The result is a permanent, man-made ecosystem that produces quality products with little maintenance. However, there is no universal definition of what a forest garden is and what not.

In Dortmund LL, the food forest was created on a 3,000 m² area of the St. Urbanus community center as a self-sustaining forest-permaculture ecosystem (Figure 2.2). An initial analysis before implementation showed that available city-owned land was either already planned for other uses or affected by heavy contamination from industrial use in the past. Therefore, the search was extended to privately owned land as well as to smaller areas.

NBS for social innovation 11

Figure 2.2 Community-based urban farms in St. Urbanus community center (NBS 3).
Source: ProGIreg

The process of realization of the urban forest took place as follows.

- Early in 2019, the nonprofit association "die Urbanisten e.V." contacted the St. Urbanus parish. The proposal to build a food forest together with the members of the parish was positively received, also since the parish already had planned to create a fruit garden at their site. After a few meetings, the board of the parish agreed to clear part of its church garden of dense vegetation.
- In May 2019, an action day took place with around 70 people participating. Five raised beds were built and planted with bee-friendly flowering plants and vegetables. The proGIreg team used this kick-off activity to capture initial ideas and gained insight into group dynamics within the community.
- Several co-design workshops with the scouts and the community council followed during summer. Wishes (e.g. planting tunnel, compost) were recorded, and framework conditions (boundaries of the garden, contamination check) were noted and a first plan drawn. This plan was further

developed together with a permaculture expert. Local conditions such as previous vegetation, shading of parts of the area by trees, or the soil conditions were regarded in the concept.
- The area was divided into zones (berry bushes, wild fruit, fruit trees, flowering plants, climbing plants, sour bed, and wilderness). Large parts of the ground were additionally covered with plantings of wild herbs and vegetables (Figure 2.3).
- Since autumn 2019, the area was gradually supplied with woodchips from a horticultural company. They were spread over the entire area, thereby serving as a protection against weeds and fertilizing the garden in the long term. The woodchips slowly compost and form a valuable humus layer, like in a forest.
- In March 2020, the final plan was presented at the community center as well as during a church event with about 100 guests. The implementation was supposed to take place immediately afterward, but the COVID-19 lockdown was a hard break in the project planning.

Figure 2.3 Schematic design of the community-based urban farms in St. Urbanus (NBS 3).

Source: die Urbanisten, Mandy Schreiber

- In May 2020, community members began to work in compliance with contact and the physical distancing rules. For this purpose, "die Urbanisten" recorded an instruction video for the soil preparation work. At the beginning, the work was done in groups of two. After the restrictions were relaxed, up to ten volunteers spread the fertilizer, covered the area with cardboard, and then covered it with woodchips.
- The newly formed gardening group met every Saturday until autumn 2020 to push ahead that first plantings could be realized.
- In June 2021, an online event took place with the aim of public relations and garden planning, focusing on plant selection and timing. Throughout summer, a series of workshops were conducted to construct a composting area, a planting tunnel, and various climbing aids. Participants also continued to plant herbs and berry bushes, with plans to dibble wild fruit bushes and fruit trees in fall (Figure 2.3). The garden group, consisting of approximately 20 individuals, coordinated its activities through a chat group, with additional workshops organized by the proGIreg project team.

Figure 2.4 Planting activities with local citizens of Huckarde in St. Urbanus (NBS 3).

Source: die Urbanisten

Figure 2.5 Planting activities with the scouts in St. Urbanus (NBS 3).

Source: Mais Jafari

14 Nature-Based Solutions for Urban Renewal in Post-Industrial Cities

- Since the beginning of 2022, the garden's citizens group met on a weekly basis adding more plants and new soil as well as spreading woodchips among the plants.

Starting in 2023, the garden citizen group has been meeting regularly to maintain the garden, with fewer meetings during winter months. In spring, the group dibbled more plants in the garden. In 2023, the partner organization, "die Urbanisten", regarded their implementation support within the proGIreg framework as completed. Subsequently, they officially transferred the management to the garden group. Figure 2.6 illustrates the co-creation cycle of NBS 3.

NBS 8: Pollinator biodiversity

To increase pollinator biodiversity in Huckarde (NBS8), urban spaces are converted into habitats that are attractive for pollinating insects. This goal can

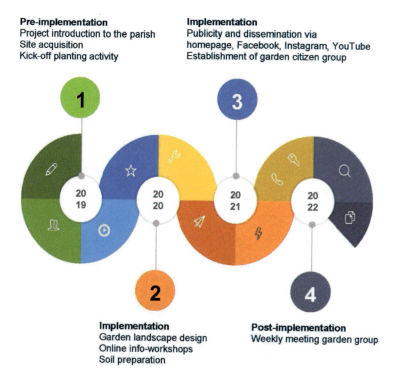

Figure 2.6 Co-creation cycle of NBS3.

Source: Authors

be achieved by establishing plants that serve as habitats themselves and those that offer rich food sources like pollen and nectar for the desired species.

While it requires taxonomic skills to design the space with perennial insect-friendly bushes and trees, for example, the early flowering Cornelian cherry tree or the ever flowering almond-leaved willow (*Salix trianda semperflorens*), oftentimes it is easier to establish flower meadows on smaller spaces. When maintained to be free of bushes, these spaces can be easily converted back and repurposed for other usages later on. The relatively low planning effort and costs, little implementation work, and the potentially temporary nature of this method make it attractive for most of the relevant stakeholders. In general, real estate developers do not have to make a long-term decision for the space, neighboring residents can enjoy the beautiful view, and the implementing group does not need an abundance of knowledge for the implementation.

Whereas in other proGIreg cities available spaces for NBS implementation could be provided right from the start, in Dortmund LL all projects started from scratch. Therefore, identification of project sites was the first challenging implementation task which turned out to be more difficult than assumed. Several housing companies were addressed but hesitant to offer their lawn areas for project ideas that still had to be defined during a co-design-process. A one-hectare large area on renatured Deusenberg landfill was offered by the owner. During a site visit with nature and insect experts the area turned out to be of high ecological value so that implementing a flower meadow would cause more damage than benefit.

Figure 2.7 Expert site visit to identify potential sites for NBS8 (Biodiversity).

Source: Mais Jafari

16 Nature-Based Solutions for Urban Renewal in Post-Industrial Cities

In the later course of this excursion, an expert from the Department of Green Spaces in the City of Dortmund offered to guide the group through Huckarde district to identify suitable public spaces for implementing NBS8. The group agreed on a few alternative spaces managed by the City of Dortmund and an allotment garden association.

Encouraged by the success of the initial site visits, networking with relevant stakeholders continued and consideration was given to how all these stakeholders and available resources could outlast the proGIreg project. While for NBS8 it is relatively easy to reach a consensus between the multiple stakeholders, it can be observed that it is the multi-stakeholder nature itself that acts as an implementation barrier. Real estate developers are potentially open to make the space available, but they are usually not the driving force behind the actual implementation. Likewise, individuals and companies might be willing to finance seeds and materials, but they lack a network of activists. Experts have the knowledge but do not have enough helping hands to push for concrete implementations of flower meadows.

Therefore, creating an association was envisioned as a social innovation. Inspiration was found within the local proGIreg consortium, as the Dortmund LL partner "aquaponik manufaktur GmbH" is also active in the field of biodiversity and had founded an association in its hometown Issum for this purpose. While discussing the concept, goals, and ideas of this association, it was concluded that they easily could be transferred to the city of Dortmund. Thus, the idea of founding "Naturfelder Dortmund" was initiated. It acts like a catalyst within the stakeholder network and helps to implement biodiversity projects whenever manpower and knowledge are missing resources. In fall and winter 2020/21, a media campaign was organized and interested citizens were invited to participate in the process. The NBS partners ("die Urbanisten e.V.", "Naturfelder Issum e.V.", supported by the City of Dortmund) used their already established social media channels (mainly their websites, Instagram, and Twitter) to inform the general public about the new initiative and to invite potential collaborators to take part in meetings. Flyers were printed on seed papers that sprout flowers when discarded and dropped on sufficiently wet soil. The idea and the call for participation have been communicated to potential participants in online presentations during conferences in sustainability and participative city planning topics.

Starting early 2021, the association members met via Zoom on a biweekly basis and discussed next steps with the participants. The newly invited participants themselves were animated to develop the statutes and to take charge of the approval process of the association. Minutes were taken and kept with an intranet system called Confluence, whereof die Urbanisten has

NBS for social innovation 17

Figure 2.8 Information booth of "Naturfelder Dortmund" at a biodiversity event.
Source: "Naturfelder Dortmund e.V."

a community license. An alternative software for this purpose might be the free and Open Source XWiki. The association "Naturfelder Dortmund e.V." was finally founded in July 2021.

"Naturfelder Dortmund e.V." took initiative even before its official establishment and created its first flowering meadow in spring 2021, utilizing an area of the "Emschergenossenschaft" which is responsible for the renaturation and maintenance of the Emscher river. The activity strictly followed the guidelines for containing the COVID-19 pandemic, ensuring compliance. In addition, a video was recorded to promote public relations efforts. Flowering meadows have been successfully sown in various private and public areas. Furthermore, the Green Space Department of the City of Dortmund received valuable support from the association by ensuring irrigation for the sown meadows; without this offer these areas would not have been seeded. In 2022, the Naturfelder group organized workshops in an elementary school, focusing on sowing a flowering meadow and creating a sandarium specifically designed for ground-living wild bees (Figure 2.8).

Social innovation aspects on implementation of community-based urban farms and gardens and biodiversity projects

In the following, factors that contributed to the social innovation and success of the project will be described.

- Urban community gardens, such as the St. Urbanus Food Forest, provide a fertile environment for cooperation, participation, and empowerment. Collaborative efforts within these gardens foster opportunities for meaningful interaction and knowledge exchange. Through the co-design process, participants gain practical experience with democratic negotiation processes, idea generation, and planning, while also acquiring new skills. Some participants come with extensive prior knowledge, while others with little gardening experience but may possess other useful skills, such as handicrafts. In addition to the knowledge experience along that process, the group also benefited from the proGIreg supervision.
- The cooperation between "die Urbanisten e.V." as a nonprofit, nongovernmental organization and the parish of St. Urbanus proved to be fruitful. After the parish council decided to jointly implement the project, all aspects of the project were supported. Appointments were arranged, the garden was promoted at parish festivals, and other citizens were invited to participate. This included, for example, a local beekeeper who was allowed to set up his hives on the grounds of the food forest. Materials such as the pallets for the raised beds were procured through the local network. Working with the local tree pruning company kept costs down and the prunings contain much more carbon than comparable material from the hardware store. Having the Scouts, who played an active role in the co-design process and construction of the garden, provided recourse of an already organized group. This facilitated the start of the project.
- Realizing the garden on church-owned land instead of urban city-owned land simplified the implementation process by easing bureaucratic obstacles. When using city-owned land it is necessary to sign a use-contract with the city administration. Tenants need to agree to regard possible use limitations which may hinder an open and collaborative co-design process. Moreover, such contracts are likely subject to time limitations and may require provisions for deconstruction at the end of the lease time. In the current scenario, these procedures were omitted by leveraging church land, allowing for a more streamlined approach to establishing the garden without the associated complexities of urban land ownership.
- Regular meetings were crucial for the success of community garden projects, particularly during the gardening season. These gatherings provided a platform for discussions, agreements, and effective planning. Through collaborative efforts, the group developed a sense of unity and shared responsibility for the management of their communal garden. Organizing educational and planning events during winter further enhanced group cohesion and ensured continuity throughout the project.

NBS for social innovation 19

Table 2.1 Participation events NBS3 and NBS8

Event	Description
Meetings, workshops	Face-to-face meetings with site owners and potential partners.
Site visits	Site visits with local experts to select and evaluate potential sites for NBS8.
Planting campaign	Co-design planting campaign to activate the various stakeholders and local community in designing, implementing, and managing NBS3 and NBS8.
Lectures and public events	Informing the general public and students about both NBS and their implemented projects in Dortmund, e.g. lecture in the church community about urban gardening and biodiversity in the annual summer school "Students for Future".
Social media, website	Information about proGIreg and the LL area. The Department of Urban Renewal in Dortmund presents the project on its homepage (www.proGIreg.dortmund.de). The different co-design activities are also covered on the web page of "die Urbanisten" (www.dieurbanisten.de). Recently, "die Urbanisten" initiated creating a website for proGIreg in Huckarde presenting the Dortmund NBS addressing Huckarde citizens and involving them in local projects activities (www.hansagruen.de). It includes project activities and implementation steps.
ProGIreg publications	Published and printed materials (press release, leaflet, roll-up, etc.).
Networking and collaborations	Networking with projects that share similar goals to exchange information of best practices, such as "nordwärts – Dortmund Kooplab".

20 Nature-Based Solutions for Urban Renewal in Post-Industrial Cities

Barriers, challenges, and opportunities: NBS3 and NBS8

Space scarcity is a major obstacle to urban agriculture and biodiversity projects in Dortmund LL: Urban space is a contested space where urban agriculture competes with other more demanding and competing interests, such as dense urban development, which often take precedence. Land ownership is another significant challenge, leading to unforeseen delays in implementation. Originally, the NBS projects were planned for areas other than the current sites, but administrative obstacles led to changes during implementation. This caused the planning process to take longer than anticipated, as evidenced by NBS3, where only one site was available for implementation despite several potential sites being identified. Negotiations with landowners were often unsuccessful, resulting in the need to find alternative sites for the projects.

The process-oriented approach of co-design can be challenging to communicate: Property owners tend to prefer clear plans and perspectives for their properties. This can complicate negotiations ahead and during the implementation of NBS projects. In addition, the lack of incentives for participation and uncertainties about long-term outcomes can make some landowners hesitant to engage in the process. The level of appreciation for the intangible outcomes of these projects also varies among stakeholders. Effective communication, clear planning, addressing concerns, and providing incentives are crucial for involving property owners and stakeholders in NBS projects.

The COVID-19 pandemic has posed significant challenges to the implementation of all NBS in Dortmund: The pandemic has not only made it difficult for people to gather physically but also has posed challenges for promoting NBS on-site. While digital events have become a common alternative, they cannot fully replace the benefits of in-person interactions. The situation has been particularly challenging for NBS3, where some elderly members had to navigate digital platforms for the first time in their lives to attend virtual events. This required them to learn how to use online meeting platforms, which may have been a barrier for some to participate. Regrettably, some individual persons were excluded from the online events due to limited access or unfamiliarity with digital technologies. Moreover, the effect of "zoom fatigue" was evident, as participants often had to spend several hours video conferencing at work before joining the biweekly digital meetings. Despite these limitations, the "Naturfelder Dortmund" association maintained their biweekly digital meetings and supplemented them with physical meetings (in compliance with the COVID-19 restrictions) to address the challenges posed by the pandemic.

Building trust within a project group is much more challenging in digital means of communication compared to physical meetings: The absence of nonverbal cues and casual conversations made it harder to establish interpersonal connections and foster trust among participants. In-person meetings provide opportunities for spontaneous interactions that contribute to getting to know and trust each other. The lack of personal interaction can impede effective collaboration and communication or make them more difficult within the project group, as trust plays a crucial role in building strong relationships and achieving shared project goals.

Co-design processes of the City of Dortmund traditionally do not start from scratch: Implementation of co-design projects usually start with securing (at least the first) space and checking site limitations before actively involving citizens. This approach is primarily driven by the need to ease project starts thus animating citizens to participate in realizable projects. For public entities like the city administration building trust and reliability with its citizens is crucial for a constructive collaboration which is needed in the long run. In comparison, nongovernmental organizations have a different status within society and more freedom to experiment with co-design projects (Figure 2.9).

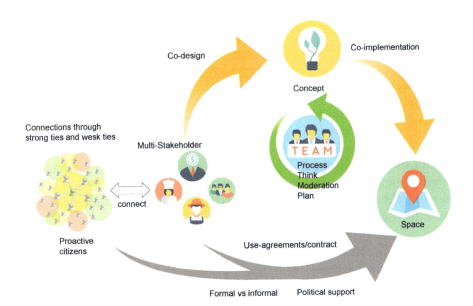

Figure 2.9 Co-creation cycle in Dortmund NBS.
Source: R. Morgenstern & M. Jafari

22 Nature-Based Solutions for Urban Renewal in Post-Industrial Cities

Administrative barriers can significantly impede project progress: Despite the expertise and experience of "die Urbanisten" and the association "Naturfelder Issum e.V." in preparing the necessary paperwork and association charter, the "Naturfelder Dortmund" encountered an obstacle when the local tax office rejected their application for the tax privilege "for the common good" due to a minor formality. As a result, the group was unable to open a bank account, which prevented them from advertising for new members or actively seeking new projects. This setback caused a substantial loss of momentum for the group, leading to a standstill in their activities for almost six months. However, the determination of the group prevailed, and they continued to pursue their goals. Eventually, the issue was resolved and the nonprofit status of the association was legally registered in August 2022.

Success factors and takeaways of social innovation in Dortmund Living Lab

In the Dortmund LL, risk-taking in social innovation of NBS has been an experimental process with many helpful outcomes and know-how for new NBS implementation that can be replicated in other contexts. By embracing a culture of experimentation and learning, the Dortmund LL has been able to advance the understanding and application of NBS, paving the way for further innovation and replication in other local and international contexts.

Openness and trust in people participating are crucial success factors: When establishing a citizen garden group or a new association, such as in NBS3 or NBS8, it is vital to prioritize authenticity and openness with other participants, ensuring transparency about institutional or project background. This includes being receptive to constructive criticism and being willing to adapt the original plan if necessary. Allowing participants to connect, define their roles, and contribute to the group's governance is crucial for success. Trusting that participants will actively engage and form an internal organization fosters ownership and empowerment. Early delegation of tasks and responsibilities frees up time and resources for community growth. This approach is similar to gardening projects, where the emphasis is not only on planting plants, but also to care for the soil and create fertile conditions. In this way, the group will naturally bloom and prosper.

Turning restrictions into new options: The COVID-19 pandemic has highlighted the importance of finding innovative ways to engage all members of the community, including those who may face difficulties with digital platforms. As the situation evolved and restrictions were lifted, it was crucial to strike a balance between digital and in-person activities to ensure inclusivity and equal participation in NBS initiatives. Digital meetings are well suited for short consultations in a small team. For the acquisition of new members or complex planning processes, physical meetings are much more effective.

References

Ayob, N., Teasdale, S. & Fagan, K. (2016). How Social Innovation "Came to Be": Tracing the Evolution of a Contested Concept. *Journal of Social Policy*, vol. 45, no. 4, pp. 635–653. https://doi.org/10.1017/S004727941600009X.

Brondizio, E., Diaz, S., Settele, J. & Ngo, H.T. (2019). Report of the Plenary of the Intergovernmental Science-Policy Platform on Biodiversity and Ecosystem Services on the Work of Its Seventh Session. Summary for policymakers of the global assessment report on biodiversity and ecosystem services of the Inter-governmental Science-Policy Platform on Biodiversity and Ecosystem Services.

Chambers, J.M., Wyborn, C., Ryan, M.E., Reid, R.S., Riechers, M., Serban, A., et al. (2021). Six Modes of Co-production for Sustainability. *Nature Sustainability*. https://doi.org/10.1038/s41893-021-00755-x.

European Commission (2015). Towards an EU Research and Innovation Policy Agenda for Nature-based Solutions & Re-naturing Cities. Brussels.

European Commission: Directorate-General for Research and Innovation (2021). *Evaluating the impact of nature-based solutions – A handbook for practitioners*. Publications Office of the European Union. https://data.europa.eu/doi/10.2777/244577.

Herlyn, E. (2021). Naturbasierte Lösungen ± aktuelle Herausforderungen und zukünftige Potenziale. In F.-T. Gottwald, J. Plagge & F. J. Radermacher (Hrsg.), *Klimapositive Landwirtschaft: Mehr Wohlstand durch naturbasierte Lösungen* (S. 13–21). Baden-Baden: Tectum Wirtschaftsverlag.

Voorberg, W.H., Bekkers, V.J.J.M. & Tummers, L.G. (2015). A Systematic Review of Co-Creation and Co-Production: Embarking on the Social Innovation Journey. Public Management Review, vol. 17, no. 9, pp. 1333–1357. https://doi.org/10.1080/14719037.2014.930505.

NBS for circular economy   **3**

Baselines for sustainable NBS implementation

Axel Störzner, Kimberly Schnell and Theresa Kemeny

Introduction

In recent years, it has become more evident that the current economic system cannot sustain long-term. Kate Raworth is explaining this in her introduction of the well-known TedX presentation about the doughnut economies model in describing the contradiction of limitless growth in the current linear economic system in a limited world. Climate change and global warming have been found to be directly connected to the current economic exploitation of the planet. Capitalism is built on the assumption that planet earth can provide infinite resources, when in fact the planet can only provide and reproduce what lies within its boundaries. Exploitation and overuse of planetary resources have caused biodiversity loss, mass extinction and the collapse of entire ecosystems. To visualize planetary limits, the planetary-boundary framework was developed (Rockström et al., 2009). The framework identifies nine boundaries within which human activity can remain secure, sustainable and long-lasting. However, if boundaries get crossed, human and planetary risks arise which cause a high degree of uncertainty and occurrence of risks. Figure 3.1 outlines the nine planetary boundaries. It shows that, by the year 2023, six of these nine boundaries are in critical status.

Building on the concept of planetary boundaries, Kate Raworth developed the model of a doughnut economy by adding social boundaries

DOI: 10.4324/9781003474869-3

This chapter has been made available under a CC-BY license.

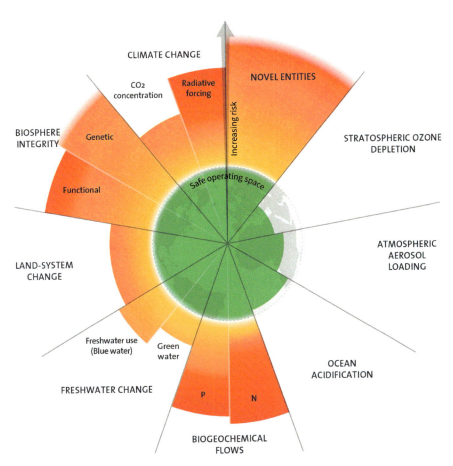

Figure 3.1 Planetary Boundaries. Credit: Azote for Stockholm Resilience Centre, Stockholm University. Based on Richardson et al. (2023), Steffen et al. (2015) and Rockström et al. (2009).

to the framework (Raworth, 2012). Illustrated in Figure 3.2, the social dimensions are building the inner circle whereas the environmental ceiling marks the outer circle boundary. Only when human activity remains in the green doughnut-shaped area, it is sustainable and lasting.

Introduction to the circular economy

Climate change advancing and the accumulation of plastic waste in the natural environment posing serious harm to global wellbeing, new schools of

26 Nature-Based Solutions for Urban Renewal in Post-Industrial Cities

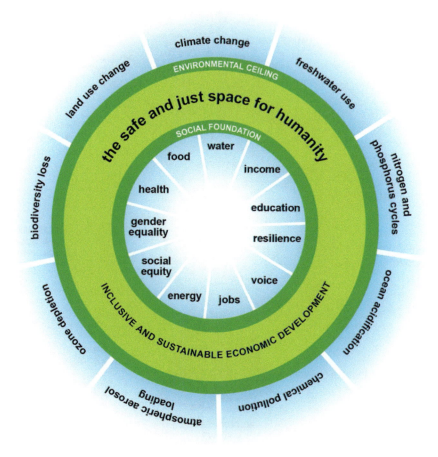

Figure 3.2 Doughnut economy (Raworth, 2012).

thought have put different ideas forward – one of them being the concept of a circular economy. Here, resources are not only used once and disposed at the end of the product's life but also re-introduced into the production phase, creating a circular flow of materials (Ellen MacArthur Foundation, 2021). Therefore, the economy moves away from the "take-make-waste" approach and to a more circular one (Ellen MacArthur Foundation, 2021). It is built on three main principles: eliminate waste and pollution, circulate products and lastly, materials (at their highest value) and regenerate nature (Ellen MacArthur Foundation, 2021).

The first principal – waste valorization – refers to the rethinking of product and packaging design. By designing a product in a way that it can be re-used and re-introduced into the product life cycle, waste is eliminated. This particularly applies to packaging and single-use items

NBS for circular economy 27

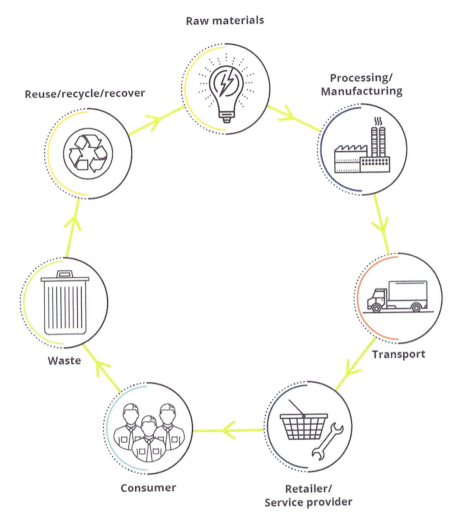

Figure 3.3 Circular economy conceptual framework.

such as disposable cutlery or to-go cups as they traditionally are designed to be thrown away after only a short utilization period. By creating circular packaging solutions, raw materials can be reused which saves resources and prevents waste. Rethinking design of a product and its packaging could also be selling products without any packaging. As the Ellen MacArthur Foundation states "The problem (and the solution) starts with design" (Ellen MacArthur Foundation, 2022). Therefore, in order to think more circularly, it is important to not only focus on developing proper recycling techniques, but think about different ways of reusing products before they are even designed.

The second principle refers to keeping products and materials in use for as long as possible and at the highest quality (Ellen MacArthur Foundation, 2022). When products cannot be used any longer, the circular approach urges to extract materials or product components that can be re-used for a new product or as a by-product. The process can be divided in two distinct cycles: the biological and the technical cycle. Whereas the technical cycle focuses on the re-usability and re-introduction of products and materials, the core of the biological cycle is the biodegradability of biological products. Both cycles aim at reducing waste and negative externalities, that is, depletion of resources, and are crucial in achieving circular economy as the new normal in industry and society.

The third principle refers to the regeneration of nature and natural capital.

> In the natural system, there is no one who takes anything out without giving it back in another form that can be further utilized. One person's waste is another person's food. If modern man intervenes in such a system, the cycle becomes a conveyor belt that only runs in one direction.
>
> (Göpel, 2020)

And most of the time at this end we found waste to be dumped.

Whereas linear economy requires constant input of new resources, circular economy makes use of materials already extracted which leaves more room for nature to recover and regenerate. Furthermore, a high focus lies on natural processes which aim at avoiding waste and emphasizing natural degradation of biodegradable materials.

Nature-based solutions for circular economy

One of the strategies to shift away from a linear economy towards a circular one is to see "nature as a source of solutions (NBS) to challenges associated with climate change" (Bourguignon Didier, 2017). Actions inspired by nature can be used in a sustainable way to solve societal, economic and environmental challenges and can facilitate the transitions towards a circular economy (Stefanakis et al., 2021). The European Commission has identified four different key areas where nature-based solutions (NBS) can be successfully integrated in the fight against climate change and shifting away from the conventional, non-sustainable system:

(i) supporting sustainable urbanization to stimulate economic growth and enhance human well-being, while making the urban area more attractive,

(ii) restoring degraded ecosystems by improving their resilience and increasing the ecosystem services they offer,
(iii) climate change adaptation and mitigation with focus on carbon storage, and
(iv) improving risk management and resilience.

(Stefanakis et al., 2021)

By implementing more NBS, the desired socio-economic development towards a circular economy can be achieved in a more sustainable way. The use of NBS methodologies and concepts supports the evaluation of the performance of different products, services and systems. This alternative approach provides a new sustainable infrastructure in a cost-effective way and simultaneously provides additional benefits and ecosystem services. However, it is necessary that policies and laws are implemented to accelerate this transition. One example is the "European Circular Economy Action Plan". Here, "NBS appear as ideal representatives of this new approach, as they not only contribute to the reduction of the carbon footprint, one of the main goals of a circular economy, but also enhance the resilience against climate change impacts" (Stefanakis et al., 2021).

Although NBS most likely will have multiple economic and climate resilience benefits, the implementation as a strategy for a climate neutral economy is lacking. Therefore, Stefanakis et colleagues (2021) highlights several steps that need to be integrated, including raising awareness, incorporating these steps into adaptation and action plans, increasing investments in nature, aligning with the European Union (EU) taxonomy, and fostering financing and cooperation, all aimed at promoting the creation of a circular economy (Stefanakis et al., 2021). The different steps are as follows:

1) Firstly, awareness must be raised on nature's value and its significance of their natural capital, resilience and benefits to humanity. For example, planting mangroves can protect coastal areas from dangerous flooding. This is evidently cheaper than artificially made barriers and nicely illustrates that nature offers a variety of solutions for problems that humanity currently faces. However, this understanding of nature must be spread through all different sectors aiming at the implementation of NBS for a circular economy in order for it to become the new normal.

2) Another important factor is that NBS should be integrated into climate adaptation plans. Since the main goal of a CE is to reduce greenhouse gas (GHG) emissions, NBS can play an important role in achieving this aim. Impact assessments on how well NBS benefit climate adaptation should be integrated in the planning, decision-making and action on

adaptation. Only by making NBS an integral part of climate adaptation, its full potential can be exploited.

3) Thirdly, investments should target NBS to foster green technologies and climate adaptation and mitigation plans (Stefanakis et al., 2021). As reported "Investments in nature based solutions (NBS) need to triple by 2030 according to the UNEP State of Finance for Nature report" (Fiona Cromarty, 2022).

4) "If the world is to meet the climate change, biodiversity, and land degradation targets, it needs to close a USD 4.1 trillion financing gap in nature by 2050. The current investments in Nature-based solutions amount to USD 133 billion – most of which comes from public sources" (Vivid Economic, 2021) (https://www.unep.org/resources/state-finance-nature).

There is great hope that the EU taxonomy will become the tool to reorient capital into sustainability and circular economy actions (source: https://finance.ec.europa.eu/sustainable-finance/tools-and-standards/eu-taxonomy-sustainable-activities_en). In the taxonomy, the circular economy is listed as one of the six environmental goals, which illustrates its importance. That means that in the future, reporting on sustainable investments will include disclosures on the circular economy. Investing in NBS can therefore become highly relevant in the context of the taxonomy.

5) Lastly, NBS should be a main factor to consider in financial conditions and policies. "The key is to link the current challenges with the available solutions and the existing expertise" (Stefanakis et al., 2021). Global financial institutions should integrate the NBS approach into their financing conditions while involving citizens and companies to actively participate in the solutions to maximize their impact. The cooperation of science, policy and practice is needed for a successful implementation of NBS (Stefanakis et al., 2021).

Consequently, NBS can play a significant role in creating a shift towards the circular economy. In the following part of the chapter, three NBS are highlighted which offer sustainable, circular solutions that evidently contribute to a more sustainable future.

Leisure activities and clean energy on former landfills for the circular economy

The NBS "Leisure activities and clean energy on former landfills" is the perfect example of how post-industrial areas can be repurposed. By using the area and setting up solar panels, like on Deusenberg in the Dortmund Living

Lab, to generate renewable energy, the first principle of circular economy is followed. Becoming independent from fossil fuels is the first step in the needed energy transition. Simultaneously, a park with an industrial past is transformed for leisure activities such as sports or to enjoy nature. This supports the well-being of the people around and a healthy and active lifestyle. This NBS complements science, policy and practice.

With the EU Green Deal, "renewable energy is promoted to become the new normal" (EU, 2022). Using the site to create green spaces, the NBS balances out negative effects of urbanization and the built environment (Pearlmutter et al., 2020). Synergies are detected with principle 3, regenerating nature and another NBS, new regenerated soil. This is one of the most important NBS for circular economy: "By moving from a take-make-waste linear economy to a circular economy, we support natural processes and leave more room for nature to thrive" (Ellen MacArthur Foundation, 2022).

Instead of importing fertile soil which led to immense environmental and financial costs, the key is to regenerate soil directly on the spot by using, for example, organic household waste to fertilize. This allows cities to thrive and nature to be built back into urban areas, providing cities with more natural capital and a higher quality of life for the population (Ellen MacArthur Foundation, 2022).

For example, green spaces lower the urban heat island effects and lower the risk of floods (proGIreg, 2022). By incorporating green and natural spaces in the built environment, carbon sinks are created which play a crucial part in reducing GHG emissions. Soil can capture immense amounts of CO_2 when being in a balanced shape and health. However, due to overuse and artificial fertilizers, most soils are currently in unhealthy conditions and are becoming carbon sources, emitting carbon to the atmosphere (Pete Smith, 2014). The NBS "new regenerated soil", implemented by an industry-partner in Turin, aims at tackling that problem. It is mainly about re-using deep excavation material (e.g. from road construction, renaturing of riverbanks, new building measures for residential and/or industry purposes) plus further "ingredients" to create a new regenerated soil to be of advantage when being used on post-industrial sites, especially parks, urban forests and urban gardening. In addition to the environmental benefits of reducing CO_2 emissions, regenerated soil also increases economic value via market-based mechanisms. Regenerating soil and grasslands can be part of an emissions trading scheme like in China (IKI, 2021) and can create revenue streams which can then be re-invested into climate change mitigation and adaptation solutions.

The Turin company even receives money for taking the excavated soil. After developing the new soil, they sell it, so that this NBS creates double-income streams: for taking a key resource and by selling the new produce.

The food system is responsible for 30% of global GHG emissions, deforestation (Crippa et al., 2021) and significant water usage. According to the UN FAO, between 2 000 and 5 000 litres of water are needed to produce a person's daily food intake (Gruere and Shigemitsu, 2021).

Therefore, it is needed that new ways of food production and innovations like community food-hubs or technical innovations find place with the result and to reduce the use of natural resources while producing high-quality foods.

Aquaponic systems are a modern way to produce fish and vegetables in a circular and sustainable way. Aquaponic systems can be explained as an efficient combination of horticulture and aquaculture. Instead of adding additional fertilizer for plants to grow, it is using fish's wastewater to provide plants with needed nutrients. Through the circular flow of water, plants are constantly irrigated and nourished (Forchino et al., 2017).

The advantages of aquaponic systems include not only the reuse of natural resources and that no soil and toxic fertilizer are needed, it also provides

Figure 3.4 A. Störzner 2019 - Aquaponic System for home use developed by Citybotanicals GmbH, Dortmund.

nutrient-rich foods for humans. In addition, it can be set up in areas which are water-scarce and do not have much arable land for agriculture. It is common that aquaponic systems are built in greenhouses to ensure favourable conditions to grow.

Therefore, aquaponic systems are in line with all three CE principles. First, the system is designed for circularity: reusing and recycling water. This goes hand in hand with principle 2 which states that in this case, water is used longer and organic matter can be degraded easily. Lastly, aquaponic systems have a positive impact on topics regarding short-distance transportation, non-use of chemical fertilizers and pesticides and depletion of natural water habitats through external fish breeding. Hence, it supports the regeneration of soil and nature which is the last principle of CE.

It is inevitable that aquaponic systems are a vital solution in the transition towards a circular economy. It is the interplay of NBS which pushes the change towards a sustainable life for all.

Critical reflection

Throughout this chapter, benefits of NBS and the circular economy have been highlighted. Yet, to provide a holistic picture of the matter, it is important to critically reflect on both: the upsides and downsides of NBS for the circular economy. Only then it is possible to reach a concise conclusion and recommendations for the future.

NBS have a very specific applicability, meaning that they require long planning periods and the manpower and knowledge of many different stakeholders. Setting up an NBS often requires a site to fit in, very specific knowledge and includes many parameters that must be considered. This is not surprising that the implementation of NBS is challenging. The costs of implementing an NBS could be rather high, while results cannot be shown in a short period of time. Change management usually requires high monetary resources and funding that oftentimes comes with commitment to deliver results at a high speed. The actual report of the UN environment programme "State of Finance for Nature" has described the pathway in three steps:

1. short-term: create a market for NBS investment
2. medium-term: support emerging markets and investment returns
3. scale-up and monitor investment

(UN Environmental Programme, 2022).

Yet, many NBS show their efficiency and positive contribution only over longer periods of time. Using new generated soil as an example, it is evident that in order to have fertile soil through the new generated soil approach, a longer period of time and many resources are required. Therefore, many different stakeholders must agree on entirely transforming waste and transportation circles in order to create a circular method to re-fertilize soil. This process cannot be implemented in a short period of time, and therefore, investments in change management are not immediately visible. That can lead to frustration and fewer support from stakeholders and reluctancy from communities to set up and support NBS solutions.

As NBS are a rather new trend within the circular economy field, not much scientific research has been conducted on the subject. Even though many of the NBS are closely monitored by scientists, the extent and timespan are not sufficient to make vast, general conclusions on NBS.

That it will play a crucial role in global transformation and nature regeneration activities is not anymore the question! This is reflected by the latest UN Biodiversity Conference (COP 15) in December 2022 in Montreal, Canada, and its high number of contributions about NBS implementation strategies.

References

Bourguignon Didier (2017). *Nature-based solutions: Concept, opportunities and challenges.* Think Tank European Parliament. https://www.europarl.europa.eu/thinktank/en/document/EPRS_BRI(2017)608796

Crippa, M., Solazzo, E., Guizzardi, D., et al. (2021). Food systems are responsible for a third of global anthropogenic GHG emissions. *Nature Food, 2,* 198–209. https://doi.org/10.1038/s43016-021-00225-9

Ellen MacArthur Foundation (2021). *Toward the circular economy.* https://www.aquafil.com/assets/uploads/ellen-macarthur-foundation.pdf

Ellen MacArthur Foundation (2022). *Regenerate nature.* https://ellenmacarthurfoundation.org/regenerate-nature

European Commission (2022). *Energy and the Green Deal: A clean energy transition.* https://commission.europa.eu/strategy-and-policy/priorities-2019-2024/european-green-deal/energy-and-green-deal_en

Fiona Cromarty (2022). *An entrepreneur's experience: The role of finance in enabling sustainable development.* UN Environment Programme. https://www.unep.org/resources/newsletter/entrepreneurs-experience-role-finance-enabling-sustainable-development

Forchino, A.A., Lourguioui, H., Brigolin, D., & Pastres, R. (2017). *Aquaponics and sustainability: The comparison of two different aquaponic techniques using the Life Cycle Assessment (LCA).* https://www.sciencedirect.com/science/article/pii/S0144860916301522

Göpel, M. (2020). *Rethinking our world: An invitation.* Ullstein Hardcover. https://doi.org/10.9783550200793

Gruere, G., & Shigemitsu, M. (2021). *Water: Key to food systems sustainability.* https://www.oecd.org/agriculture/water-food-systems-sustainability/

IKI (2021). *Nfga plans to explore the economic value of carbon sinks from forestry and grassland through the carbon market.* https://climatecooperation.cn/climate/nfga-plans-to-explore-the-economic-value-of-carbon-sinks-from-forestry-and-grassland-through-the-carbon-market/

Pearlmutter, D., Theochari, D., Nehls, T., et al. (2020). Enhancing the circular economy with nature-based solutions in the built urban environment: Green building materials, systems and sites. *Blue-Green Systems*, 2(1), 46–72.

Pete Smith (2014). *Do grasslands act as a perpetual sink for carbon?* https://onlinelibrary.wiley.com/doi/full/10.1111/gcb.12561

proGIreg (2022). *New regenerated soil.* https://progireg.eu/nature-based-solutions/new-regenerated-soil/

Raworth, K. (2012). *A safe and just space for humanity: Can we live within the doughnut?.* Oxfam.

Rockström, J., Steffen, W., Noone, K., et al. (2009). Planetary boundaries: Exploring the safe operating space for humanity. *Ecology and Society*, 14(2), 1–33.

Stefanakis, A. I., Calheiros, C. S., & Nikolaou, I. (2021). Nature-based solutions as a tool in the new circular economic model for climate change adaptation. *Circular Economy and Sustainability*, 1, 303–318.

UN Environmental Programme (2022). *Nature finance action tracks.* https://wedocs.unep.org/xmlui/bitstream/handle/20.500.11822/36151/SFN_Inf.pdf

Vivid Economics. (2021). Green Stimulus Index. An assessment of the orientation of COVID-19 stimulus in relation to climate action and biodiversity goals. https://www.klimareporter.de/images/dokumente/2021/02/GreenStimulusIndex5thEdition.pdf

NBS for biodiversity

4

*Simona Bonelli, Federica Larcher,
Lingwen Lu, Marta Depetris
and Manuela Ronci*

Introduction on nature-based solutions for biodiversity

The term 'nature-based solutions' (NBS) was introduced by MacKinnon et al. (2008) and Mittermeier et al. (2008) both focusing on the solutions to mitigate and adapt to climate change effects while simultaneously protecting biodiversity, building capacity, and fostering resilience. Beginning in the 2000s, and emerging strongly in development discourses around 2017, NBS gained ground both as a principal and an umbrella of approaches and technologies (Hanson et al., 2020). NBS is defined as actions to protect, sustainably manage, and restore natural or modified ecosystems that address societal challenges effectively and adaptively, simultaneously providing human well-being and biodiversity benefits (Cohen-Shacham et al., 2016). NBS is also defined as any behaviour using ecosystem services, which is conducive to reducing the consumption of non-renewable natural resources and increasing the protection of renewable natural resources (Maes & Jacobs, 2017). The concept of NBS has evolved into an umbrella concept that includes concepts such as green/blue/natural infrastructure, ecosystem approaches, and ecosystem services, but at their core, learning and using nature to create sustained socio-ecological systems that enhance human well-being and biodiversity protection (Dick et al., 2019).

As defined by the Secretariat of the Convention, by biodiversity we mean "the variety of life: the diversity of all living organisms from the various ecosystems of the planet. It includes diversity within species, between species and of ecosystems in which they live". Based on traditional biodiversity conservation and management strategies, NBS integrate science,

DOI: 10.4324/9781003474869-4

This chapter has been made available under a CC-BY license.

policy, and practice to create biodiversity benefits in well-managed diverse ecosystems (Eggermont et al., 2015). NBS are not a substitute for the rapid decarbonization of all sectors of the economy but can be a complementary solution to effectively address the joint challenges of climate change and biodiversity loss. To achieve this, they must be well designed, properly implemented, and efficiently managed, and longevity, target species, appropriate participatory approaches, and state of current habitat and scale need to be considered (Girardin et al., 2021). NBS currently focus on the protection of intact ecosystems, managing working lands, restoring native cover, and creating novel ecosystems in urban settings. Such activities score high on mitigation, biodiversity, and adaptation co-benefits and can be cost-effective and scalable.

The use of nature and technology can have an impact on the future supply of other services. COP26 revealed that whereas many organizations and governments are embracing the approach as an essential tool for tackling climate change, others, particularly grassroots organizations, have dismissed it as a dangerous distraction from systemic change (Melanidis & Hagerman, 2022). Nathalie Seddon (2022) suggested that NBS can make an important contribution to achieving net-zero carbon emissions this century, but only if combined with other climate solutions including slashing greenhouse gas emissions across all economic sectors. Achieving net-zero carbon emissions and transitioning to a nature-positive economy will also require systemic change in the way we behave as societies, shifting to a dominant worldview that is based on valuing quality of life and human well-being rather than material wealth – and connection with nature rather than its conquest. Signals such as the rise of climate and nature grassroots activism indicate that this shift is taking place. If carefully implemented to ensure that multiple values of the natural world are respected, NBS can offer an opportunity to accelerate this transition while also slowing warming, building resilience, and protecting biodiversity. To further understand the research themes of NBS and biodiversity, we conduct author keywords co-occurrence network analysis with the 347 publications comprised of articles and reviews searched by the keyword "nature-based solutions" or "nature-based solutions and biodiversity" from the Clarivate's Web of Science Core Collection using the software of VOSviewer. Three major themes have been identified, including ecosystem services value assessment, enhance urban ecological resilience with green infrastructure, and climate change mitigation and adaptation with biodiversity conservation.

Ecosystem services were defined in 1997 as the processes and outputs provided to human during the transformation of natural resources (Costanza et al., 1997). Deksissa et al. (2021) suggested that a city can combine the development of urban agriculture and urban green infrastructure to overcome barriers for enhancing ecosystem services. In addition, a growing body of research highlights the contribution of ecosystem services provided by urban forests to the quality of urban life (Baro et al., 2014). How to balance the needs of human beings and the ability of the earth to provide ecosystem services is considered to be one of the greatest challenges of this century. Many management strategies have applied the concept of ecosystem services to meet this challenge, including land management, policy, and economic decision-making to achieve overall ecosystem health (McDonough et al., 2017). Biodiversity plays an important role in regulating ecosystem services. One of the important functions is that biodiversity can buffer environmental changes and maintain certain ecosystem services when the ecosystem faces disturbances. Importantly, the production of most ecosystem services depends on the plants and animals in the ecosystem, although there is often no simple relationship between their quality and quantity and the diversity of wild animals and plants. Consequently, the mainstream view is that when biodiversity elements are lost, ecosystems will become less resilient (Harrison et al., 2014), thus ecosystem services value assessment has becoming an important part in the study of biodiversity.

A resilient city refers to a city that is capable of resisting disasters, reducing disaster losses, quickly recovering to a pre-disaster state, and being able to learn from past disasters and accidents to improve resilience to disasters. Among them, urban ecological resilience refers to the ability of cities to recover and adapt to natural disasters such as global warming, floods, and heavy rains. The theme of enhancing urban ecological resilience with green infrastructure is related to Sustainable Development Goal (SDG) 11, "Make cities and human settlements inclusive, safe, resilient and sustainable". Lehmann (2021) introduced the integration of NBS as a strategy in urban planning with the aim to strengthen urban ecological resilience and to slow down the biodiversity decline. In addition, NBS provide a systematic approach to promote the maintenance, enhancement and restoration of biodiversity and ecosystem services in urban areas, helping to increase the resilience of urban areas (Beceiro et al., 2022). Green infrastructure is widely recognized for reducing flood risk, improving water quality, and harvesting storm water for future us, which would be an important part of strategies used in urban planning to enhance sustainable development and urban resilience (Fu et al., 2021). Therefore, enhancing urban ecological resilience with green infrastructure would be a focus in the future research field.

Mitigation and adaptation are two equally important aspects of addressing climate change. NBS mitigation of climate change includes

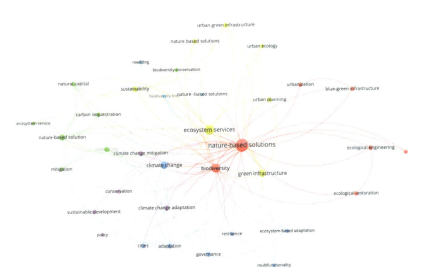

Figure 4.1 Keywords co-occurrence network analysis based on the Clarivate's Web of Science Core Collection.

three aspects: protection of natural ecosystems, restoration of natural ecosystems that have been damaged or degraded by slopes, and sustainable management of farmland, grassland and woodland. At the same time, strengthening the capacity building to adapt to climate change, especially to deal with extreme climate events, is an important guarantee for the realization of the UN SDGs. Climate resilience is an important mix of climate mitigation and climate adaptation designed to minimize current and future disruption while promoting opportunity (Beery, 2019). Particularly worth mentioning is that biodiversity conservation will be a hot research topic in this field for climate mitigation and climate adaptation towards the future climate resilience.

NBS and benefit for animal biodiversity

NBS is a relatively young concept, still in the process of being defined.

NBS is more considered as an umbrella concept that covers a range of different approaches that have a common focus on ecosystem services and aim to address societal challenges.

In the late 2000s, the term 'nature-based solutions' emerged taking into account people not only as passive beneficiaries of nature's benefits, but as active part of the process, proactively protecting, managing and restoring

40 Nature-Based Solutions for Urban Renewal in Post-Industrial Cities

natural ecosystems as a significant contribution to addressing major societal challenges (Cohen-Shacham et al., 2016).

One of the first definition of NBS was clearly referred to solutions taken to mitigate impacts that come from nature and come back to nature protecting ecosystems and biodiversity (MacKinnon et al., 2008; Mittermeier et al., 2008).

This point was better clarified some year later by the International Union for Conservation of Nature that defined NBSs as "actions to protect, sustainably manage and restore natural or modified ecosystems that address societal challenges effectively and adaptively, simultaneously providing human well-being and biodiversity benefits" (Cohen-Shacham et al., 2016).

While the definition of ecosystem services referred to an immediate benefit to human well-being and economy, NBSs include a broader spectrum of topics, focusing also on the benefits to people and the environment, in order to find sustainable solutions that are able to respond to long-term environmental changes.

In this way, the concept of NBS goes beyond the traditional definition of biodiversity conservation, as it also takes into account social issues such as human well-being and poverty alleviation, socio-economic development, and governance principles (Eggermont et al., 2015).

Nevertheless, the application of NBSs has been often concentrated in catching advantages in terms of guaranty ecosystem services to human ecosystems forgetting the biodiversity outcome. This uncorrected approach is reversed in the circular economy view, where nature is considered a model to imitate, in order to find new solutions (i.e. biomimicry, ecosystem services valuation, and bioeconomy), but there is not any feedback for biodiversity in this growing process Moreover, this kind of process could contribute to biodiversity conservation, but much more often it happens that they may also contribute to biodiversity in negative ways, especially when applications are not applied carefully (Buchmann-Duck & Beazley, 2020).

This lack is due to a confusion in the definition of NBS that in some cases included the definition provided by the European Commission in 2015, which lacks the benefit for biodiversity: Nature-based solutions aim to help societies address a variety of environmental, social and economic challenges in sustainable ways. They are actions inspired by, supported by or copied from nature; both using and enhancing existing solutions to challenges, as well as exploring more novel solutions, for example, mimicking how non-human organisms and communities cope with environmental extremes (European Commission, 2015).

Thus, recently Seddon and colleagues (2020) underlined the importance of correcting planning any NBSs following essential criteria, one of which is that the NBSs should be explicitly designed to provide measurable benefits for wild biodiversity.

In fact, contrary to many engineering solutions, NBS could be able to face the challenges of climate change while delivering multiple benefits for people and nature, at a relatively low cost.

It has been pointed out that NBSs that protect and restore natural ecosystems, using native species, can have an important role in opposing climate change, while contributing to cultural and social ecosystem services. By contrast, NBSs that do not take into account biodiversity conservation are more vulnerable to environmental changes in the long term. Designing an NBS which evaluates biodiversity conservation is an important detail that includes two processes: planning the NBS in order to enhance biodiversity and measuring the positive effects on it.

In screening literature, there are very scarce projects based on NBS that plan to give back positive effects on animal conservation.

Despite many publications linking the two words NBS and biodiversity (Figure 4.1), only few of them refer to animal diversity and most of them do not refer to study cases where NBS give an explicit and measurable impact on animal diversity (Melanidis & Hagerman, 2022).

The main habitat involved in these projects is the urban environment and agro-ecosystem, naturally poor in animal diversity, but dependent on nature for their long-term survival. Some ecosystem services such as soil quality and pollination are crucial in simplified and anthropogenic ecosystems, like crops or cities. In these systems any activity that can enhance the diversity can help in maintaining overall ecosystem services.

In 2019, for example, Catarino and colleagues consider that a crop cultivated in respect of insect pollinators could be considered as a 'nature-based agriculture' which uses ecological principles for sustainable agro-ecosystems, balancing ecology, economics, and social justice. The authors demonstrated that as happens in pollinator-dependent crops and in oilseed rape cultivation, a system based on agro-ecological principle not only can give a positive effect on insect pollinators increasing their abundance, but at the same time, brings to a reduction in the use of chemical inputs. This kind of approach can increase yields from 16% to 40% higher in fields with a high pollinator abundance. This is a clear demonstration that the promotion of NBS for agricultural production can be an alternative to conventional agriculture for both food production and farm income. Even within cities, public gardens and allotments can be pollinator hotspots if managed properly and improved by NBSs (such as vegetable gardens with nectariferous and host plants), thereby becoming shelter and food sources for these insects (Baldock et al., 2019).

As in the agro-ecosystems, NBSs have increased interest in recent years, especially in urban environments, with the aim to increase local wellness and mitigate climate change effects. Most of the works on NBSs usually refer to solutions based on the use of plants to improve human well-being, but what is missing are NBS which refer to the role of animals as a tool to increase the quality of life in cities.

Within urban ecosystems there are some applications that increasingly recognize the benefits from nature to improve human well-being, such as many initiatives that bring nature back into urban areas in order to design more biophilic cities (defined by E.O. Wilson as the innate tendency of humans to focus on life and life processes), meaning centres designed to incorporate nature into urban environment and provide close contact with nature for their citizens, who also take care of this nature (Beatley & Newman, 2013), that can give us and our pets a more comfortable lifestyle, reducing the level of stress and mental illness, focusing on the concept of 'people and nature' to build more resilient systems (Granai et al., 2022).

Despite the confusion that characterizes the definition of NBS, if we want to have the best results, we must design these solutions considering all aspects on which we are going to act, thus considering not only to imitate nature, but also to provide a benefit to it. Indeed, by bringing benefits to biodiversity and the environment, there will also be benefits in economic, social, and human health terms.

NBS, design strategies, and plant biodiversity

Two crucial aspects that have emerged so far are the complex interrelationship of environmental problems and opportunities and the need to apply a new approach in the definition of strategies that not only serve the needs of mankind but are also able to return benefits to nature. In this framework, landscape architecture proves to be a powerful tool to address both biodiversity loss and climate change through the configuration of open spaces that can accommodate heterogeneous and resilient ecosystems. The integration of NBS in the design can be particularly useful in pursuing the goal of constructing biotopes and spaces suitable for other-than-human species. Especially the use of vegetation – one of the main compositional elements of landscape design – is recognized as a major driver of biodiversity (Chong et al., 2014; Mayrand, 2020). In fact, plants are an essential biotic element for creating peculiar ecological conditions, while at the same time favouring the development of plant biodiversity and the presence of varied animal communities.

NBS for biodiversity 43

Further effective strategies relate to the post-implementation phases and are linked to the maintenance and management of the projects over time.

In terms of possible approaches to fostering biodiversity, a landscape architecture project can intervene in various ways, depending on the overall objectives and the context in which the designers operate.

One category of action consists of preserving and enhancing existing ecological conditions. These spaces can be defined as what the ecologist Ingo Kowarik (2011) called 'first nature', i.e. remnants of original ecosystems that present the characteristics of the potential vegetation of a site and that, although embedded within an anthropized matrix, have been scarcely subject to human pressure.

A second type of space for conservation is that of wilderness developed as a result of restricted access or abandonment. Typically, these are sites that are devoted to a specific use and, thus, protected by fences – such as archaeological sites and cemeteries – or decommissioned spaces – such as industrial areas and infrastructures – that, being subject to less anthropic disturbance, underwent spontaneous colonization processes by plant communities (Gandy, 2016).

While the first typology refers to the management of existing situations, a different mode of action in favour of biodiversity consists of restoring or artificially constructing precise biotopes and ecological niches, making use of different levels of engineering to provide several ecosystem services and manage coexistence between species (Salizzoni, 2021). These interventions can range from the reconstruction of riverbanks and wetlands to the reforestation of large areas, but also include the integration of green walls and roofs into architectural designs.

A selection of international examples will be briefly reviewed to illustrate these two approaches.[1] While certainly not exhaustive, the aim of the following overview is to provide some suggestions with respect to the use of NBS in the design and management of open spaces intended to promote biodiversity.

Conservation and management of existing ecological conditions

Naturpark Schöneberger Südgelände, Berlin (DE) | 1996–2009

Naturpark Schöneberger Südgelände is a public park stretching on a former brownfield where vegetation successions occurred spontaneously, in the

Schöneberg district of Berlin. The project is unanimously recognized as a successful attempt to balance a burst of undisturbed nature with the remains of human activity, reconciling wilderness conservation with the active use of the space.

The park is located on a disused shunting station, which remained isolated and undisturbed for about 50 years since the post-war division of the city, allowing for the spontaneous development of a high level of urban biodiversity (Heatherington, 2014). Thus, it was the citizens themselves who demanded that the site be converted into a park (Langer, 2012) and, after German reunification, it was eventually recognized as a nature conservation area in 1999. The project was completed in the following ten years by the state-owned company Grün Berlin.

One of the main objectives in implementing the design was to ensure accessibility to the park without damaging wilderness. The issue was solved defining a system of paths based mainly on the existing railway track, ramps, and underpasses, in order not to add further elements of fragmentation other than those already present. In protected areas established as a conservation site – such as the delicate arid grassland – access is provided via a raised metal footbridge, which allows both accessibility for people and protection of the biotopes from trampling (Figure 4.2).

Figure 4.2 A raised footbridge allows the public to observe and cross the space, while protecting biotopes from trampling. On the left side, flocks are used for grazing. Photo: Manuela Ronci, 2021.

Figure 4.3 Non-dense groups of trees, maintained as groves, thrive among the railway tracks. Photo: Manuela Ronci, 2021.

A second issue was related to the degree of intervention in spontaneous vegetation dynamics, which was addressed by defining various spaces and biotopes subject to differentiated maintenance regimes.

The spaces identified are clearings, non-dense groups of trees maintained as groves (Figure 4.3), and wild plant masses, for which the spontaneous dynamics are instead left undisturbed to develop.

Among the management strategies, it is worth mentioning that meadows are regularly mowed using flocks as a sustainable solution based on grazing (Figure 4.2).

The success of the park is certainly given by the fascinating mixture of railway wrecks, sculptures, and disruptive nature, together with an elegant design that contains construction details demonstrating a specific attention to the preservation of wilderness (Figure 4.4).

Jardin des oiseaux, Paris (FR) | 2012

Jardin des oiseaux (or Enclos des oiseaux) is a small garden located near the Père Lachaise monumental cemetery, in the 20th arrondissement of Paris. The site consists of an old wasteland (*friche*) which was transformed by the

Figure 4.4 The design of the path system was shaped and adapted according to the presence of the vegetation already developed on site. Photo: Manuela Ronci, 2021.

Municipality of Paris into a public space, following a participative process involving the local population.

Despite the size, it is a successful example of differentiated management, which allowed the garden to be awarded with the EcoJardin label[2] in 2012. At the core of the garden is a wild species meadow, mowed once a year and fenced off to make it inaccessible to the public, which can instead move along the path embracing the meadow (Figure 4.5).

The boundaries of the garden are made of scattered trees, melliferous shrubs – suitable for attracting pollinating insects – and plants producing berries, which are intended to provide shelter and resources for numerous bird species. In addition, the garden includes nesting boxes for birds, shelters for bats, and a hotel for insects (Figures 4.6 and 4.7).

Branches, dead leaves, and mowing residues are regularly kept on the site. This choice has the double effect to provide habitats for insect communities and small mammals as well as a natural soil fertilizer, when the organic materials decompose.

The existing plant biodiversity has been preserved as much as possible by recovering materials, soil, and seeds present *in situ*, providing citizens with a privileged place to observe the vegetation and wildlife.

NBS for biodiversity   47

Figure 4.5 The garden includes a central wide meadow and is enclosed by trees and masses of shrubs. Photo: Manuela Ronci, 2021.

Figure 4.6 The garden hosts different typologies of nests for wildlife. Photo: Manuela Ronci, 2021.

Figure 4.7 The garden hosts different typologies of nests for wildlife. Photo: Manuela Ronci, 2021.

Construction of biotopes and ecological niches

Parc de Billancourt, Boulogne-Billancourt (FR) | 2011–2017

Parc de Billancourt is located in the heart of Boulogne-Billancourt, a municipality west of Paris, in proximity to the river Seine. The park is the driving force behind a new urban expansion for the district and has provided for the renaturalization of a site with an industrial past.

In the intentions of the designers, Agence Ter studio, it was intended to be the fulcrum of a new innovative storm water management system, collecting water from the surrounding blocks and open spaces (Peñin & Ferrater, 2011). A further, crucial objective was to create the conditions for biodiversity to spread in the park, defining a true ecological niche in relation to the Seine and the other green spaces of the neighbourhood.

Parc de Billancourt plays a central role in filtering rainfall, acting as a lamination basin in the event of flooding, and is also designed for water storage, thanks to permeable soils, ponds (Figure 4.8), and recessed rain gardens. Water recovery (part of which is reused for irrigation) takes place in a system consisting of a marsh and a peat bog, adjacent to a beach and sand ditches. The upper parts of the park are covered by a grassland that is not very demanding in terms of water consumption.

This concept based on rational water management involves a great variety of plants and ecosystems, as well as spatial heterogeneity and differentiated maintenance. The northern part of the park, which is the most exposed to sunlight, contains dry herbs and large, open lawns suitable for outdoor recreation (Figure 4.9).

In contrast, the shadier part of the park is home to wetlands with high ecological value and is intended for passive recreation and staying.

These two portions are separated by an intermediate band which can be flooded (Figure 4.10). This section is lower than the level of the park and has a changing appearance due to variable moisture and seasonal conditions.

Figure 4.8 Water feature at Parc de Billancourt, a project designed to act as a lamination basin and a reservoir for urban biodiversity. Photo: Manuela Ronci, 2021.

Figure 4.9 The park includes large lawns and dry herbs with law water requirements. Photo: Manuela Ronci, 2021.

Figure 4.10 Shaping the topography of the site, the designers defined lower, temporarily floodable areas, that contribute to the ecological heterogeneity of the park. Photo: Manuela Ronci, 2021.

The botanical choice was mainly oriented towards the selection of numerous native species, aiming to guarantee the sustainability of the ecosystems with a view to ecological management of the space. Thanks to the robust focus on the environmentally sustainable design and management, this park was also awarded the EcoJardin label.

Green walls in Piazzale Aldo Moro, Turin (IT) | 2020–2022

The fourth case refers to the application of NBS in an architectural framework. The green walls of Piazzale Aldo Moro, in the centre of Turin, were implemented on two buildings of the Aldo Moro complex (Figure 4.11), a new university campus commissioned by Università degli Studi di Torino and entrusted to architect Claudio Bobbio.

The project achieves benefits related to thermal and acoustic comfort, through the cooling and insulation of the buildings, but also to the enhancement of plant diversity in a dense urban fabric.

Nearly 30 species of plants were selected and aggregated, according to ornamental and optimal growth criteria, to create the vertical greenery. The solution was realized using a structure covered with a waterproofed wooden panel, appropriately spaced from the buildings to ensure ventilation. The species were arranged in felt pockets attached to the structure and interwoven with the automated irrigation system. The fabric, capable of retaining water, allows water loss to be limited and maintains a level of humidity adequate for the plants' survival.

Figure 4.11 The green walls enrich the site with environmental benefits as well as ornamental effect in a densely built fabric. Photo: Manuela Ronci, 2022.

The project made use of herbaceous perennials and small shrubs, which enrich the plant stock of a heavily sealed area. As part of the site regeneration process, other micro-interventions were carried out to re-vegetate the site as much as possible, making use of climbing plants (Figure 4.12) or building small gardens (Figure 4.13).

The value of small-scale interventions such as the two green walls in Turin is linked to replicability and the possibility of developing a widespread system of micro-spaces capable of supporting biodiversity even in densely built-up urban areas.

Conclusions

Design can intervene in a robust way to implement plant biodiversity through NBS. Solutions can be varied, involving large urban parks, but also residual spaces and buildings. The value of these interventions is particularly high in urban contexts, where it is possible to intervene to strengthen

Figure 4.12 Other small interventions (such as the use of climbing plants, on the right) contribute to enhance the presence and diversity of plant species in the neighbourhood. Photo: Manuela Ronci, 2022.

NBS for biodiversity 53

Figure 4.13 The green wall constitutes an iconic tile of the green infrastructure that has been developed. On the background, the trees of 'Il bosco degli altri', a garden recently developed on the site by Lineeverdi firm. Photo: Manuela Ronci, 2022.

existing conditions or create opportunities for the development of plant communities in heavily sealed and built-up areas.

In this sense, even small-scale interventions can contribute to the implementation of plant diversity, and their role becomes crucial, especially in dense urban fabric and historical centres, where the space available for the survival of plant species is often very small.

The tools that landscape architecture can use include both design strategies and maintenance scheduling. While the former is intended as a means of constructing different spatial and ecological conditions, the latter is a fundamental factor in the preservation of these desired conditions and, therefore, in ensuring that the project responds effectively to the objectives of enhancing biodiversity.

Although the literature still shows a strong lack of evidence for the beneficial effects of NBS on fauna, strengthening plant biological diversity might turn out to be a powerful strategy to contribute to the enhancement of animal diversity as well, since plant species are the basic element for the

54 Nature-Based Solutions for Urban Renewal in Post-Industrial Cities

configuration of ecosystems and ecological niches capable of hosting animal communities.

Notes

1. Along with the support of literature sources, the descriptions are the result of field surveys conducted in 2021 and 2022.
2. EcoJardin label is the highest standard in the ecological management of open spaces, coordinated by Agence Régionale de la Biodiversité (Regional Agency of Biodiversity).

References

Baldock, K. C., Goddard, M. A., Hicks, D. M., Kunin, W. E., Mitschunas, N., Morse, H., … & Memmott, J. (2019). A Systems Approach Reveals Urban Pollinator Hotspots and Conservation Opportunities. *Nature Ecology & Evolution*, 3(3), 363–373.

Baro, F., Chaparro, L., Gomez-Baggethun, E., Langemeyer, J., Nowak, D. J., & Terradas, J. (2014). Contribution of Ecosystem Services to Air Quality and Climate Change Mitigation Policies: The Case of Urban Forests in Barcelona, Spain. *Ambio*, 43, 466–479.

Beatley, T., & Newman, P. (2013). Biophilic Cities Are Sustainable, Resilient Cities. *Sustainability*, 5(8), 3328–3345.

Beceiro, P., Brito, R.S., & Galvao, A. (2022). Assessment of the Contribution of Nature-Based Solutions (NBS) to Urban Resilience: Application to the Case Study of Porto. *Ecological Engineering*, 175, 17.

Beery, T. (2019). Exploring the Role of Outdoor Recreation to Contribute to Urban Climate Resilience. *Sustainability*, 11 (22), 6268. https://doi.org/10.3390/su11226268.

Buchmann-Duck, J., & Beazley, K.F. (2020). An Urgent Call for Circular Economy Advocates to Acknowledge Its Limitations in Conserving Biodiversity. *Science of the Total Environment*, 727, 138602.

Catarino, R., Bretagnolle, V., Perrot, T., Vialloux, F., & Gaba, S. (2019). A Nature-based Solution in Practice: Ecological and Economic Modelling Shows Pollinators Outperform Agrochemicals in Oilseed Crop Production. *BioRxiv*, 628123.

Chong, K.Y., Siyang, S., Kurukulasuriya, B., Chung, Y.F., Rajathurai, S., & Tiang Wah Tan, H. (2014). Not All Green Is as Good: Different Effects of the Natural and Cultivated Components of Urban Vegetation on Bird and Butterfly Diversity. *Biological Conservation*, 171, 299–309.

Cohen-Shacham, E., Walters, G., Janzen, C., & Maginnis, S. (2016). *Nature-based Solutions to Address Global Societal Challenges*. Gland, Switzerland: IUCN.

Costanza, R., D'Arge, R., & Groot, R.D. (1997). The Value of the World's Ecosystem Services and Natural Capital. *Nature*, 4, 253–260.

Deksissa, T., Trobman, H., Zendehdel, K., & Azam, H. (2021). Integrating Urban Agriculture and Stormwater Management in a Circular Economy to Enhance Ecosystem Services: Connecting the Dots. *Sustainability*, 13, 19.

Dick, J., Miller, J.D., Carruthers-Jones, J., Dobel, A. J., Carver, S., Garbutt, A., ... & Quinn, M. (2019). How Are Nature Based Solutions Contributing to Priority Societal Challenges Surrounding Human Well-being in the United Kingdom: A Systematic Map Protocol. *Environmental Evidence*, 8, 11.

Eggermont, H., Balian, E., Azevedo, J. M. N., Beumer, V., Brodin, T., Claudet, J., ... & Le Roux, X. (2015). Nature-based Solutions: New Influence for Environmental Management and Research in Europe. *Gaia-Ecological Perspectives for Science and Society*, 24, 243–248.

European Commission, Directorate-General for Research and Innovation (2015). *Towards an EU Research and Innovation Policy Agenda for Nature-Based Solutions & Re-naturing cities: Final Report of the Horizon 2020 Expert Group on 'Nature-based Solutions and Re-naturing Cities': (full version)*, Publications Office. https://data.europa.eu/doi/10.2777/479582

Fu, X., Hopton, M.E., & Wang, X.H. (2021). Assessment of Green Infrastructure Performance through an Urban Resilience Lens. *Journal of Cleaner Production*, 289, 11.

Gandy, M. (2016). Unintentional Landscapes. *Landscape Research*, 41(4), 433–440.

Girardin, C.A.J., Jenkins, S., Seddon, N., Allen, M., Lewis, S.L., Wheeler, C.E., ... & Malhi, Y. (2021). Nature-based Solutions Can Help Cool the Planet—If We Act Now. *Nature*, 593(7858), 191–194. https://doi.org/10.1038/d41586-021-01241-2

Granai, G., Borrelli, C., Moruzzo, R., Rovai, M., Riccioli, F., Mariti, C., ... & Di Iacovo, F. (2022). Between Participatory Approaches and Politics, Promoting Social Innovation in Smart Cities: Building a Hum–Animal Smart City in Lucca. *Sustainability*, 14(13), 7956.

Hanson, H.I., Wickenberg, B., & Alkan Olsson, J. (2020). Working on the Boundaries—How Do Science Use and Interpret the Nature-based Solution Concept? *Land Use Policy*, 90, 104302.

Harrison, P.A., Berry, P.M., Simpson, G., Haslett, J.R., Blicharska, M., Bucur, M., ... & Turkelboom, F. (2014). Linkages between Biodiversity Attributes and Ecosystem Services: A Systematic Review. *Ecosystem Services*, 9, 191–203.

Heatherington, C. (2014). Buried Narratives. In A. Jorgensen & R. Keenan (Eds.), *Urban Wildscapes*. Abingdon and New York: Routledge, 171–186.

Kowarik, I. (2011). Novel Urban Ecosystems, Biodiversity, and Conservation. *Environmental Pollution*, 159, 1974–1983.

Langer, A. (2012). Pure Urban Nature: Nature-Park Südgelände, Berlin. In A. Jorgensen & R. Keenan (Eds.), *Urban Wildscapes*. Abingdon and New York: Routledge, 152–159.

Lehmann, S. (2021). Growing Biodiverse Urban Futures: Renaturalization and Rewilding as Strategies to Strengthen Urban Resilience. *Sustainability, 13*, 21.

Mackinnon, K., Sobrevila, C., & Hickey, V. (2008). *Biodiversity, Climate Change and Adaptation: Nature-based Solutions from the World Bank Portfolio*. Washington: C World Bank.

Maes, J., & Jacobs, S. (2017). Nature-Based Solutions for Europe's Sustainable Development. *Conservation Letters, 10*, 121–124.

Mayrand, F. (2020). Du choix des espèces végétales. In P. Clergeau (Ed.), *Urbanisme et biodiversité. Vers un paysage vivant structurant le projet urbain*. Rennes: Éditions Apogée, 106–116.

Mcdonough, K., Moore, T., & Hutchinson, S. (2017). Understanding the Relation-ship between Stormwater Control Measures and Ecosystem Services in an Urban Watershed. *Journal of Water Resources Planning and Management, 143*, 9.

Melanidis, M.S., & Hagerman, S. (2022). Competing Narratives of Nature-based Solutions: Leveraging the Power of Nature or Dangerous Distraction?. *Environmental Science & Policy, 132*, 273–281.

Mittermeier, R.A., Totten, M., Pennypacker, L.L., Boltz, F., Mittermeier, C.G., ... & Langrand, O. (2008). *A Climate for Life: Meeting the Global Challenge*. Arlington, VA: International League of Conservation Photographers.

Peñin, A., & Ferrater, C. (2011). 7ha Boulogne-Billancourt: le parc Billancourt. In H. Bava, M. Hössler, & O. Philippe (Eds.), *357,824 ha de paysages habites par l'Agence Ter*. Paris, Bruxelles: AAM Editions, 126–141.

Salizzoni, E. (2021). Progettare la distanza: interazioni uomo-natura nei nuovi ecosistemi urbani. In A. Gabbianelli, B.M. Rinaldi, & E. Salizzoni (Eds.), *Nature in città. Biodiversità e progetto di paesaggio*. Bologna: Il Mulino, 103–119.

Seddon, N. (2022). Harnessing the Potential of Nature-based Solutions for Miti-gating and Adapting to Climate Change. *Science, 376*, 1410–1416.

Seddon, N., Chausson, A., Berry, P., Girardin, C.A., Smith, A., & Turner, B. (2020). Understanding the Value and Limits of Nature-based Solutions to Climate Change and Other Global Challenges. *Philosophical Transactions of the Royal Society B, 375*(1794), 20190120.

Dortmund Living Lab 5

Margot Olbertz and Mais Jafari

The Dortmund Living Lab as testing ground for nature-based solutions

Living Labs (LL) are increasingly popular in cities as an innovative tool to test, validate and develop innovative solutions such as nature-based solutions (NBS) in real-life settings where citizens and a network of stakeholders are systematically involved from the early stages in co-design and co-implementation processes. This chapter presents the spatial, social and sustainable development of the LL in the district Huckarde in Dortmund, one of the three European Front Runner Cities (FRC) in proGIreg.

Dortmund (c. 609,000 inhabitants, 2022) is a typical post-industrial city in the heart of the former coal mining and steel manufacturing Ruhr metropolis area in Germany. The industry's decline in the 1970s left the city to transform economically, socially, and environmentally. Large-scale contaminated brownfields, former industrial and transport infrastructures required redevelopment and social problems had to be addressed. The necessity of developing nature-oriented solutions and improving the green infrastructure (GI) system in the city of Dortmund, and in particular in the post-industrial northern area along the Emscher River, has been at the centre of the formal and informal planning agenda in Dortmund since the early 1990s. This process is going to continue in the coming decades. Hence, proGIreg's nature-based urban renewal approach fits with the city's strategic planning framework to improve simultaneously the social, economic and environmental qualities of the urban regeneration efforts in the LL and beyond.

DOI: 10.4324/9781003474869-5

This chapter has been made available under a CC-BY license.

58 Nature-Based Solutions for Urban Renewal in Post-Industrial Cities

The overarching goal of the Dortmund LL is developing a systemic GI network by improving connectivity therein and thus enhancing the living and environmental conditions in Huckarde. The co-developed, context-specific NBS implemented in the LL aim at improving the quality of life of the local communities by offering attractive and diverse open urban spaces, and boosting collaborative and long-lasting engagement including vulnerable and marginalized groups. This in turn provides health, environmental and economic benefits as well as social inclusion and cohesion in this socially polarized part of Dortmund. Long term, the goal is to disseminate and replicate NBS and practices at other locations in Dortmund, and national and international cities.

No real-life laboratory can avoid unforeseen obstacles and different types of barriers, for example, technical/technological, administrative, social and financial. Given the post-industrial past, soil contamination poses major challenges, in particular impacting urban agriculture/food production, which is highlighted in the case study of NBS 4, community-led aquaponics. Dispersed and private landownership of envisaged spaces for NBS interventions also led to significant changes to initial plans and caused delays, exacerbated by the COVID-19 pandemic limiting active stakeholder engagement in NBS co-creation activities.

The Living Lab in post-industrial district Dortmund Huckarde

The Dortmund LL is located in the Huckarde district in the post-industrial heart of Dortmund and is characterized by densely built-up areas and large post-industrial sites. Deindustrialization from the 1960s onwards (over 10,000 workers lost their jobs within few years) has driven Huckarde into structural change with tremendous economic, social and environmental effects.

The area faces multiple regeneration challenges in regard to environmental degradation, socio-economic disparities and lack of quality green spaces and corridors, for example, poor spatial accessibility to rest of the city and to the nearby re-natured former landfill Deusenberg serving as a multi-functional recreation area, social segregation and low levels of job opportunities (see also Table 5.1).

The Dortmund LL area runs along the Emscher river next to the Huckarde district, stretching from the West of the city centre (2 km from the city centre) to the former coking plant Hansa and the former Deusenberg landfill site in the North (Figure 5.1). The Dortmund analysis area (2,275 ha) comprises adjacent districts in a 500 to 2,000 m wide buffer around the LL.

Dortmund Living Lab 59

Table 5.1 Key characteristics of the Dortmund Living Lab in Huckarde

Key characteristics	Description/challenges
Size	215 ha
i) Population	c. 40,000
Environment	ii) Lack quality green and open spaces
Socio-economic profile	Highest unemployment rates, incl. highest youth-unemployment within the city, especially in Mailoh subdistrict of Huckarde

Figure 5.1 View of Huckarde and Dortmund city from Deuseberg.

Source: Mais Jafari, City of Dortmund

To stimulate urban regeneration, the LL aims at experimenting with innovative solutions using nature and natural processes to transform brownfields and underused spaces together with the local community and diverse stakeholder groups. Turning weaknesses into opportunities represent strategic starting points for developing and implementing several NBS in the LL over the project duration. Implemented NBS will be

60 Nature-Based Solutions for Urban Renewal in Post-Industrial Cities

maintained by local citizens, non-governmental organizations (NGOs) and the local authorities beyond project end.

The effects of implemented NBS may have a direct impact on the analysis area as numerous inhabitants are living in several settlement areas directly adjacent to the LL: Huckarde in the North-West, Deusen in the North-East, Dorstfeld in the South-West, the Rheinische Straße quarter and the Union quarter in the South. The total number of residents in the analysis area is 56,812 (Figure 5.2). This is relevant for performing district-level analysis and NBS benefit assessment where the LL boundaries soften.

Policy frameworks impacting the Dortmund Living Lab

LL are ideally interconnected with the wider scope of urban development plans to leverage potential synergies and cross-fertilization between programmes for an integrated urban planning approach.

In response to the industrial decline, regeneration programmes kick-started in 1992 to strengthen the Huckarde district as a liveable neighbourhood. Various and partly overlapping green and blue infrastructure planning initiatives in the Dortmund LL exist to develop the area around the former Hansa coking plant and the Deusenberg. Therefore, it was necessary to explore urban renewal policy frameworks to leverage synergies with proGIreg NBS for an integrated urban planning approach. The city of Dortmund has implemented several integrated formal and informal planning instruments at the regional, city and district levels with the goal of improving the social, economic and environmental qualities of the regeneration area through implementing NBS in the Deusenberg, Hansa and Huckarde areas in the post-industrial north of Dortmund in general, and in Huckarde in particular.

- At the forefront is the local initiative "Nordwärts"[1] (2015–2025), raising public awareness of the strengths of Dortmund's northern districts. Numerous projects bind tradition with modernity and the industrial mining history with future-oriented living spaces in various dialogue and participation processes such as biodiversity, social resilience and public art projects.[2]
- The large-scale International Garden Exhibition Ruhr 2027 (RVR Ruhr, n.d.) aims at developing high-quality open space around the Hansa coking plant.
- The Integrated Action Concept "Huckarde-Nord" was launched in 2016 as part of the urban renewal project, and several measures are being implemented to improve connectivity through GI and redevelopment of the "Hansa Revier Huckarde" (Hansa coking plant, Deusenberg and Mooskamp light railway museum) in Huckarde.

Dortmund Living Lab 61

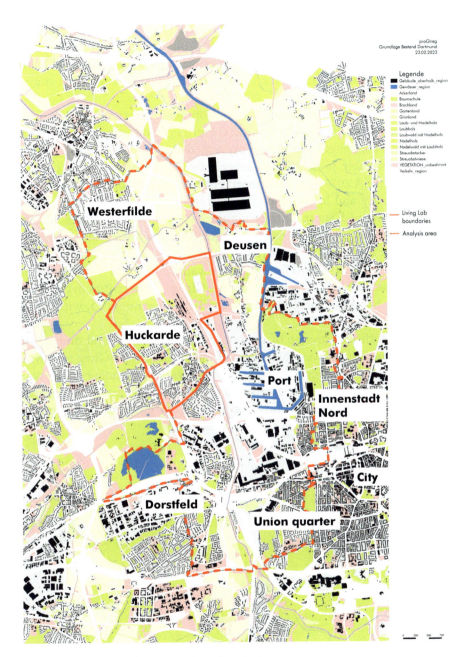

Figure 5.2 Dortmund Living Lab and analysis area.
Source: City of Dortmund

The Dortmund LL aligned its initial boundaries (215 ha) along the Emscher river with above-mentioned development programmes to leverage potential synergies with existing projects to improve the quality of life, health and well-being of citizens. However, these needed adjusting during the course of the project.

Despite obvious synergies with development programmes to sustain and manage proGIreg's NBS beyond the project's duration, a number of unforeseen challenges limited or prevented the implementation of certain NBS. Notably, joint developments between the International Garden Exhibition (IGA 2027) and NBS interventions proved difficult given IGA's higher priority and longer time horizon, for example, the initially identified and planned Sports Infrastructure NBS at Deusenberg landfill site had to give way to IGA plans. This required searching for an alternative location. In addition, a new urban renewal project being implemented on the former HSP site (Hoesch Spundwand und Profil GmbH) in the southern part of the LL excluded any proGIreg NBS activities in this area. Figure 5.3 shows the shared boundaries of the Dortmund LL with development projects.

Despite sufficient overlap of the conceptual framework and time frame of development initiatives, the degree of integration of proGIreg NBS with other regeneration programmes in Huckarde varies depending on the type of implemented NBS in the Dortmund LL, for example, proGIreg's goal of fostering biodiversity is currently not the focus of Huckarde development

Boundaries proGIreg 2018- Part of the 215 ha 'Emscher Northwards' area

Boundaries International Garden Exhibition-IGA 2027

Boundaries proGIreg 2021

Figure 5.3 The Dortmund Living Lab in relation to the boundaries of Nordwärts and IGA 2027.

Source: City of Dortmund and proGIreg

programmes, as most current initiatives aim to promote economic, social and well-being benefits in Huckarde, thus proGIreg complements existing urban regeneration strategies.

Implemented NBS in the Dortmund Living Lab

Out of eight NBS tested in proGIreg, five NBS were implemented in the Dortmund LL (Figure 5.4). Each implemented NBS addresses a set of challenges with the aim to achieve context-specific (urban) sustainability that integrates urban vitality, ecological responsibility, economic prosperity and social justice as its foundation by involving local communities.

The NBS comprise leisure activities and renewable energy production on former landfills (NBS 1), urban food production (NBS 3), aquaponics (NBS 4), accessible green corridors (NBS 6) and enhancing biodiversity (NBS 8). The NBS are inspired and supported by nature and natural processes, and complement existing and planned formal urban plans and informal initiatives in the LL and neighbouring areas. Each implemented NBS addresses a set of challenges with the aim to foster context-specific (urban) sustainability and increase urban vitality, ecological responsibility, economic prosperity and social justice and cohesion.

Table 5.2 provides an overview of NBS activities in the Dortmund LL:

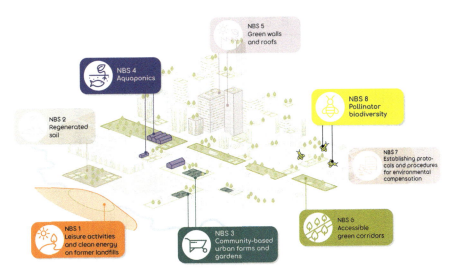

Figure 5.4 Five of proGIreg NBS implemented in the Dortmund Living Lab.
Source: RWTH

Table 5.2 Overview of NBS activities in the Dortmund Living Lab

Type of NBS	Description	Implementation
NBS 1: Leisure activities and clean energy on former landfills	NBS 1.1: Integrating solar energy production on Deusenberg landfill	Already established before the proGIreg project start
	NBS 1.2: Exercise park in Huckarde	Installation of sport devices in a public park (Gustav-Heinemann Park) usable by Huckarde and Dortmund citizens and pupils from adjacent school
NBS 3: Community-based urban farms and gardens	iii) Food forest and permaculture orchard in Huckarde	A group of engaged boy and girl scouts from the beginning of the process, together with the local NGO "The Urbanisten" and elderly church members cleared an overgrown unused space owned by the catholic parish of Huckarde. Other local residents have joined growing food, fostering social interaction in the neighbourhood
NBS 4: Aquaponics	Community managed aquaponics system	Soil contamination issues required a lengthy planning process – see case study
NBS 6: Accessible green corridors	Connection of Huckarde Borough with the re-natured Emscher river and Deusenberg sites	115 m barrier-free foot and bike path to connect the existing Emscher river path with a paved maintenance road south of Deusenberg landfill
NBS 8: Pollinator biodiversity	Improving and monitoring pollinator biodiversity in conjunction with NBS 3	Supporting pollinator diversity and biodiversity through planting pollinator friendly plant species, either on its own or in combination with other NBS activities. Formerly intensively mowed public areas or underused spaces have been transformed into flowering pollinator-friendly and species-rich meadows offering valuable habitats for pollinators. Active citizen engagement through the establishment of the citizen-led association "Naturfelder e. V.", which promotes urban biodiversity in the LL and the whole city and facilitates citizen participation in biodiversity projects.

Integration of implemented NBS into the existing policy framework

The two implemented NBS, NBS 1 Sports Infrastructure and NBS 6 Connecting Huckarde with Deusenberg and Emscher, are well integrated into the framework of the "Integrated Action Concept Huckarde-North". NBS1 is in line with the concept of creating a larger sports space network and is located on the north-south connection between Rahmer Wald via the allotment association (Hans auf Glück) to the exercise park.

Biodiversity in Huckarde (NBS 8) was initially implemented in a few locations, but then evolved into a social initiative by establishing a citizen-led association with the goal of raising citizen awareness of biodiversity and providing expertise to those who want to convert their land into biodiversity areas. The association assists with expertise and activities regarding land preparation and seeding. As this NBS is suitable for temporary conversion of small-scale areas and private land, it does not clash larger-scale redevelopment programmes.

In contrast, the incremental approach to NBS 4, community-based aquaponics was based on learning by doing in many phases, hence requiring several attempts to integrate it with other projects. Eventually, the city and IGA representatives agreed to continue operating the aquaponic greenhouses at the IGA site until 2027, being the first of its kind in Dortmund.

Stakeholder collaboration in the Living Lab Dortmund

Testing NBS in the Dortmund LL required the joint efforts of diverse stakeholders to make it happen. The complex organizational, administrative and legal aspects and overcoming financial and implementation constraints required intensive communication with all stakeholders: these included Huckarde citizens, the city of Dortmund, an active neighbourhood NGO and other city-wide operating initiatives to local SMEs and the university.

Stakeholder constellations vary by NBS depending on specific know-how or connections of key stakeholders. Identifying key stakeholders and assigning primary responsibilities and roles to manage and maintain each NBS during planning, co-designing, co-implementing and co-maintenance phases required a number of meetings and participatory workshops in the initial LL setup stages. While the Department of Urban Renewal of the city of Dortmund plays a key role in enabling urban sustainability transition processes and facilitates administrative procedures across municipal departments, active citizen involvement and other local stakeholders characterize collaborations in the LL.

66 Nature-Based Solutions for Urban Renewal in Post-Industrial Cities

Living Lab Map Dortmund Huckarde

The Living Lab Map (Figure 5.5) locates the five implemented NBS within the boundaries of the LL area in Huckarde and illustrates the NBS network the interventions create to strengthen the GI system within the LL. The map was conceived as a living document to highlight the different stages of development of each planned NBS interventions (i.e. ideas for future, in planning, in progress, implemented) and is regularly updated throughout the project to track progress.

The map is an important communication and dissemination tool, providing a concise and visual overview of the LL activities including short descriptions of each NBS, photos and the partners and stakeholders involved while indicating further potential to foster nature-based urban regeneration. It was used in participatory co-design workshops to communicate the benefits of using nature for urban renewal at concrete spatialized examples to the local community, business and industry entrepreneurs and other organizations active in the district. Not least it fostered inter-departmental understanding of the goals of proGIreg NBS within the municipality of Dortmund.

Challenges and opportunities of implementing NBS

Urban regeneration in post-industrial cities faces a number of major challenges. Transforming former industrial sites and brownfields touches on a number of complex problems. Table 5.3 provides an overview of overarching barriers to establish NBS in the Dortmund LL:

Searching for suitable NBS sites in the LL proved challenging unless land is publicly available or owned by the municipality. Many private landowners negotiate terms upon clear plans and time frames of actions. Hence, two diametrical approaches clashed. The process-oriented co-design approach applied in proGIreg proved difficult to communicate when engaging with local citizens, delaying planning and participatory processes. However, severe technological barriers hampered NBS design and implementation more significantly.

The map of soil contaminated sites (Figure 5.6) demonstrates the high degree of contamination in the city of Dortmund, ranging from fillings, industrial waste, coal mines and the coking plant to landfills as a legacy of its industrial past. These conditions are particularly prominent in the LL area (Figure 5.6), causing delays and significant additional cost of around 40% on top of the implementation budget for two NBS in the Dortmund

Dortmund Living Lab 67

Figure 5.5 Dortmund Living Lab map locating and describing NBS interventions.
Source: ProGIreg, Lohrberg Stadtlandschaftsarchitektur, 2023

Table 5.3 Overview of encountered types of barriers by NBS

Type of barrier	Description	NBS mostly impacted
Administrative/institutional	• Bureaucracy • Lengthy municipal processes • Landownership issues	Aquaponics Green corridors Leisure facilities on former landfill
Technological	• Soil contamination, pollution • Lack of expertise, knowledge, skills • Long-term maintenance	Aquaponics Leisure facilities on former landfill
Financial	• Limited budget • Additional budget required • Long-term maintenance	Aquaponics Green corridors
Social/cultural	• Limited community engagement	Aquaponics Green corridors

Figure 5.6 Soil contamination map for Dortmund, Germany – pink marks contaminated sites.

Source: City of Dortmund

LL, for example, constructing the aquaponics greenhouses on the former coking plant site and the path connecting Huckarde with the former landfill site Deusenberg.

Soil contamination limits the potential for food production, thus a number of envisaged sites for urban gardening activities (NBS 3: Community gardens and urban farms) had to be abandoned such as a planned urban garden in a public park close to housing estates in Huckarde. Establishing an aquaponics system on the former coking plant required serious adaptations to comply with food safety regulations (see case study) and substantial investments to prevent any harmful emissions.

To establish NBS 6: Enhancing the green corridor between the Huckarde neighbourhood and the Deusenberg former landfill by creating a barrier-free footpath, two soil experts had to be hired who investigated inhomogeneous landfill filling, and on top warfare objects of the Second World War was found, jeopardizing health protection and safety regulations. This resulted in lengthy planning permission procedures over a time span of three years and incurred one-third extra cost of the overall budget allocated to implement the NBS.

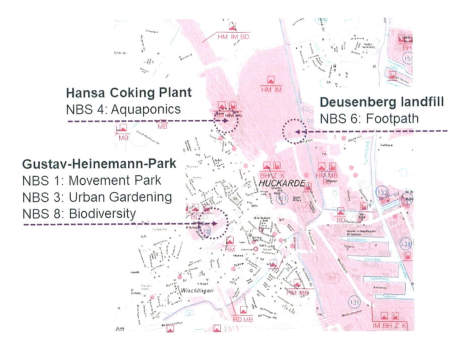

Figure 5.7 Soil contamination in the Dortmund Living Lab, Huckarde, impacting on NBS implementation.

Source: City of Dortmund

70 Nature-Based Solutions for Urban Renewal in Post-Industrial Cities

Many of these barriers incurred additional costs unforeseen at the start of the project. Finding solutions to overcome barriers demands a great deal of flexibility to change/adapt plans, gathering knowledge about procedures, work steps, time frames and distribute responsibilities. From the beginning, NBS in potentially contaminated areas should include contingency planning for both envisaged time frame and further finances/funding possibilities that can be drawn from when needed.

Impacts of the COVID-19 pandemic on Living Lab Dortmund

In light of the COVID-19 pandemic, many planned stakeholder events had to be postponed or cancelled, hence delaying a number of NBS in its planning and implementation stages. However, new methods and alternative formats to physical meetings, such as online meetings, dissemination of NBS activities and upcoming events via the proGIreg Huckarde website and blog and tutorial videos on implementation processes shifted the degree and spectrum of interaction between citizens of the LL, universities, NGOs and the municipality to a large extent to digital media. Despite its limitations, it proved a key tool to reaching out to all stakeholders and disseminating project activities. As the mode of communication changed significantly, it was interesting to observe that the acceptance of proGIreg and the willingness to participate in the co-creation process occurred regardless of virtual or a real environment communication format. Continuous and clear communication about the different NBS objectives and progress as well as mutual benefits of all groups involved proved to be a critical success factor in gaining trust while maintaining the momentum in the project implementation.

Case study: aquaponics system in the Dortmund Living Lab

The implementation of the aquaponics system serves as an example of a dynamic, non-linear NBS co-creation process. The introduction of aquaponics in Dortmund was envisioned by proGIreg given its potential to grow food on polluted post-industrial land with poor soil quality. The case study highlights the number of administrative, financial and environmental barriers that can occur and how to overcome them.

Aquaponics is a sustainable food production system that combines fish and plant production using aquaculture and hydroponic systems, with aquaculture fish process water providing nutrients for the plants. The economic performance of aquaponics facilities is subject to strong scaling effects, and sustainable business models could be developed for small to medium facility sizes, which at best also allow interim use of already planned brownfield sites for several years (Morgenstern et al., 2016). Therefore, constructing two identical greenhouses on part of the former Hansa coking plant site in Dortmund Huckarde will allow producing plants while technologically developing the concept of aquaponics. The aquaponics system will serve as a learning venue for workshops with students and citizens of the Huckarde district in Dortmund. The grown products are not marketed as in conventional aquaponics models, the NBS experiments with renting hydroponic beds to interested citizens (rent-a-raft inspired by the established rent-a-field concept). Such a model may attract citizens interested in sustainable own food production but are not willing or able to cultivate a classic allotment garden for cost reasons or required workload. It also offers potential to intensify urban–rural connections. The aquaponics production system can be the central location of a marketplace where farmers of the surrounding area market other products. Concepts such as solidarity-based agriculture and market hawkers feature established business models.

Given the complexity of the aquaponics system, the implementation process was a learning experience for all stakeholders. Overcoming numerous challenges and formal requirements required collaboration across an interdisciplinary team and the city administration. However, significant changes to the concept and operating model emerged during the NBS implementation phases described as follows in Table 5.4:

Transferring knowledge from the Dortmund Living Lab in other contexts

Communities and municipalities can use LLs as open demonstration spaces for replicating and upscaling NBS in other urban regeneration contexts. The tested NBS in the Dortmund LL have improved the GI network and social cohesion in the district. It also helped create awareness about green issues and raise interest to get engaged.

Project partners continue to work in the fields of aquaponics, urban gardening and urban renewal, thus building on networks and knowledge gained during proGIreg that impacts beyond the LL boundaries. Replication potential has been identified to extend networks regarding aquaponic systems at schools, new green corridors on contaminated sites, ground preparations to connect Hansa coking plant and the Deusenberg landfill.

Table 5.4 Challenges and barriers and solutions during co-creation phases

Phase	Challenge/barrier	Solution
Pre-implementation	Site search and identification of project location	• City of Dortmund suggested the site at the Hansa coking plant at an early stage, given its unrivalled location in the LL and the prospects of being a future hotspot for the International Garden Exhibition IGA 2027. • The site's historical significance as an industrial heritage ensures high visitor numbers, hence offering maximum public visibility. • Also serving as a demonstrator of the potential of implementing NBS on post-industrial sites. • Slightly delayed, the lease agreement between the South Westfalia University of Applied Sciences and IDS (Kokerei Hansa) was signed in February 2020.
	Site owner (IDS) demanded a deposit of 10.000€, which the designated tenant of the aquaponics site "Die Urbanisten" was not able to cover	• Extended negotiations and a time-consuming process of developing alternative solutions, South-Westphalia University of Applied Sciences (SWUAS) became official tenant, thus solving the stalemate. • In 2019, the city of Dortmund took a political resolution to pay the deposit in case SWUAS would not be able to sign the contract.
	Obtaining building permit	• Obtaining approval for constructing the greenhouses required a building application submitted to the Building Regulations Office of the city of Dortmund. • Draft application submitted in June 2020 for preliminary review and feedback, • Final draft officially submitted in November 2020, requiring major changes to the concept and operating model of the aquaponics system.

Dortmund Living Lab **73**

	The aquaponics building application was not approved for use by the general public, prohibiting co-design activities. Structural design of the foil greenhouse certified for agricultural uses only. Otherwise requiring additional calculations of the building's structure and load-bearing capacity	• Proposing new solutions to adapt the "rent-a-raft" concept, allowing citizens to rent microgarden units for producing own food. • The Urbanisten may offer visitor tours of the facility, excluding visits during extreme weather conditions. • In addition, organising workshops outside the greenhouse and at venues at Hansa coking plant.
	Gaseous emissions from the contaminated soil at the aquaponics site required additional food analysis for harmful substances	• Lab analysis to detect harmful substances. Initially done by a certified lab while SWUAS develops capabilities for these types of analyses. This will allow for continuous monitoring of the food produced at a lower cost. • In case of finding elevated values, analysis results will be corroborated by the certified lab.
Implementation	Increasing construction costs	• Soil contamination cost accumulated to 40% of overall implementation budget • Construction of contamination proof foundation constituted the most costly item
Post-implementation	Extended implementation time of 15 months (time loss and personal resources)	• Operation to be continued by proGIreg partner SWUAS after the end of the project until the end of IGA 2027.

74 Nature-Based Solutions for Urban Renewal in Post-Industrial Cities

Figure 5.8 Dortmund Living Lab aquaponics system site preparation.
Source: Die Urbanisten

Figure 5.9 Dortmund Living Lab aquaponics greenhouses construction.
Source: Die Urbanisten

Numerous urban gardening initiatives are active in Dortmund. First replication efforts by proGIreg and the city of Dortmund show effect to join forces in gathering relevant stakeholders, that is, private landowners and NGOs working with marginalized groups to seek synergies in further establishing urban gardens. Some implemented NBS offer significant potential to be

Figure 5.10 Dortmund Living Lab two identical aquaponics greenhouses.
Source: Margot Olbertz

transferred and applied in other contexts, such as neighbouring districts to harness the NBS benefits:

- NBS 1, the exercise park, is conceived as part of a network of playground and sports infrastructure in Huckarde North as part of the Huckarde open space development programme "Freiraumkonzept Huckarde" by the city of Dortmund, Department of Urban Renewal. Yet transferability of this NBS to other contexts requires substantial financial and manpower resources involving technical construction documents and tenders for the procurement of equipment by the public authorities.
- NBS 3 – Community-based urban farms and gardens at St Urbanus church is a unique community project with social impact in Huckarde district. It is maintained by the local community garden group and has a significant replication potential. Working with private landowners circumvents much bureaucracy and has proven successful for urban gardening projects. The project partner Die Urbanisten is collaborating with other urban gardening projects in Dortmund (e.g., small aquaponics projects, building insect hotels and raised beds)

76 Nature-Based Solutions for Urban Renewal in Post-Industrial Cities

- NBS 4 – Community-led aquaponics will be further developed and operated on the site of the Hansa coking plant as part of the IGA until 2027 as an innovative form of urban agriculture. Moreover, SWUAS has integrated it into an international research project INCiTiS-Food (2023–2026) with international partner universities to investigate the potential of aquaponics as an alternative for small-scale farmers to produce food in developing countries.
- NBS 6 – Footpath connecting Huckarde with the Deusenberg will be part of the barrier-free access to the International Garden Exhibition Ruhr: 2027 (RVR Ruhr, n.d.).
- NBS 8 – Enhancing pollinator diversity has gained significant interest from other parts of Dortmund due to its low-tech, low-bureaucracy application, and being led by the local community in more grassroots movements.

Post-industrial urban renewal is particularly prone to unforeseen challenges, that is, NBS implementations on contaminated soil require significant extra planning time, expertise and finances. Experimenting with establishing NBS in the Dortmund LL pinpointed to the need to intensify integrated and inter-departmental collaboration urban regeneration planning to navigate complex regulatory requirements that foster a solution-oriented approach.

The LL provided valuable testing ground and impulses for urban transformation using NBS to strengthen the GI network. However, given the predominately small-scale proGIreg NBS, notable effects on the local economy or solving structural disbalances in Huckarde requires continued urban regeneration efforts embedded in other urban planning strategies for Huckarde and Dortmund in general.

Notes

1. Project "Nordwärts", https://dortmund-nordwaerts.de/.
2. Dortmund Nordwärts projects https://dortmund-nordwaerts.de/projekte/.

References

Morgenstern, R., Biernatzki, R., Boelhauve, M., Braun, J., Dapprich, P., Gerlach, A., Haberlah-Korr, V., Mergenthaler, M., Mistele, B., Schuster, C., Winkler, P., Wittmann, M., & Lorleberg, W. (2016). Pilot study "Sustainable aquaponics production for North Rhine-Westphalia". Series of research reports from the

Department of Agricultural Economics Soest and the Institute for Green Technology and Rural Development No. 43, Department of Agricultural Economics at the University of Applied Sciences Südwestfalen, Soest.

RVR Ruhr. (n.d.). *International garden exhibition IGA 2027*. RVR Ruhr. Retrieved January 14, 2025, from https://www.rvr.ruhr/en/topics/ecology-environment/international-garten-exhibition-iga-2027/#:~:text=For%20the%20first%20time%2C%20the,urban%20landspaces%20in%20metropolitan%20regions

Zagreb Living Lab 6

*Bojan Baletić, Iva Bedenko
and Marijo Spajić*

The Zagreb Living Lab – building communities through nature-based solutions

Zagreb (c. 770,000, 2022) is the administrative, cultural, and economic center of Croatia. Like many other European cities, it went through a transition from an industrial to a post-industrial city. This transition was happening in the context of change in Croatia from a socialist planned economy to a market economy. The result of this process are numerous brownfield locations in and around Zagreb. In the city of Zagreb there are more than 40 brownfield locations, some smaller in size and some larger, with importance for development of entire city districts. One of these sites is the Sljeme brownfield area in Sesvete. The transformation of Sljeme redefines the whole district of Sesvete, and positions it as the eastern center of Zagreb. The first step in this complex and long-lasting process is to promote and operationalize the Zagreb Living Lab (LL) where new urban interventions can be developed and analyzed, co-designed with the local stakeholders, promoted and showcased for citizens of the district and Zagreb.

Sesvete is a city district on the eastern administrative edge of Zagreb, situated between the Northern foothill of the Medvednica mountain and the Sava river plain in the South. The center of Sesvete lies at the contact between the hills and the plain. This area is crossed by numerous traffic roads and the railway that links Zagreb with the main European traffic corridors. With the population of 74,000 inhabitants, Sesvete community is the biggest city district, characterized by the highest population growth due to immigration and natural population growth. On average, it is the

DOI: 10.4324/9781003474869-6

This chapter has been made available under a CC-BY license.

youngest community in Zagreb. As a community it is rather traditional, very tightly connected, with an entrepreneurial mindset; but a community that is adamant in its intentions to make Sesvete a better place to live.

Living Lab in Zagreb's Sesvete neighborhood

In the center of Sesvete neighborhood lies the abandoned meat industry complex of Sljeme. It originated from a small private enterprise established in 1879, which moved to its present location in 1921 and after the Second World War became an important socialist company that generated the economic growth of Sesvete. Its heyday was in the 1960s when a pig farm was developed to the south of the site, at the time one of the biggest in the world. The slow decline of Sljeme started in the 1990s and the industry was closed in 2006. As generations worked in the company and families' incomes depended on it, at the time the closure presented a crisis for the community.

Following the closure, the pig farm was redeveloped as a housing project for 10,000 inhabitants. The industrial part, with its 125,000 m² area, became one of Zagreb's brownfields waiting for an urban regeneration initiative. Presently, some spaces in Sljeme are rented out and parts of the open space are used as a dumping ground for used city equipment (lamp posts, kiosks, etc.). Part of the Sljeme area is unsafe due to the bad condition of some of the buildings and the danger of pieces of material falling from the buildings. The entire area is pedestrian unfriendly – there is no sidewalk and no shops or restaurants, or public parks.

In 2015, the local NGO from Sesvete invited the Zagreb Faculty of Architecture to propose a vision for change. The resulting study "Green and Blue Sesvete" offered a planning response to local needs and a long-term roadmap for urban regeneration. According to the study, the future urban development should be based on green principles and sustainable lifestyles that include green transportation, digital entrepreneurialism, a circular economy approach, and the benefits of share culture. With this aim, the area of Sljeme would form a LL for Sesvete while acting as a catalyst for future developments in the entire city of Zagreb. This was a base for Zagreb's participation in the proGIreg project. The proGIreg project is the first step in reclaiming this area by the local people to utilize it for public needs and as a common public space.

In Zagreb, proGIreg project activities are managed by the city of Zagreb as the local coordinator, in cooperation with the local partners (Zagreb Faculty of Architecture, Bureau of Physical Planning, and Green and Blue

Figure 6.1 Former Sljeme meat factory is the location for Zagreb Living Lab.

Sesvete NGO), who contribute to various aspects of implementations in the LL area. The project strives to ensure participation of all relevant stakeholders in the co-design process to make it inclusive while addressing the needs of as many potential users as possible.

Policy frameworks impacting the Zagreb Living Lab

Sesvete's urban development is defined by the general urban plan (GUP Sesvete) that has not changed significantly in the past few decades in spite of critical changes in local economy and society. Although the population in the quarter of Sesvete reached nearly 80,000, some of the public facilities and services are the same as when it was far smaller. Some ten years ago the local citizens gathered around the local NGO started demanding adequate public facilities, green urban spaces, a main square, better public facilities, an efficient road network, bike lanes, a secure crossing of the railway, new space for the music school, and an innovative and creative hub. And most of all a new and clear form and identity.

These aspirations had been articulated in the study called "The Green and Blue Sesvete" (2016) made by the Faculty of Architecture in Zagreb. The community demanded from the Zagreb mayor, and his successor, an openness to change and rethink the GUP. The city's government declared its commitment to develop the post-industrial area of Sljeme to create public facilities needed by Sesvete residents. Unfortunately, little has been

Figure 6.2 View from the Living Lab across Sesvete district and its road and rail infrastructures.

done in the past eight years and all the changes in the GUP have been small interventions. Those that were important for the reuse of some buildings for the Zagreb LL activities were never carried out.

Another planning document that is important to the Zagreb LL is the urban development plan for the Sesvete North economic zone. It defines a production/commercial zone in the center of the neighborhood, with minimal percentage of other uses. It is transforming the existing green areas into brown and dividing the living areas of Sesvete.

The study "Green and Blue Sesvete" proposed to remove the present centrally positioned economic zone to a new area 5 km to the east. In its place it defines a new urban mixed-use area for 20.000+ inhabitants, connecting the two existing parts of Sesvete with a wide green corridor. As the ideal starting point for this new Sesvete downtown development serves the area of the former meat industry complex Sljeme, property of the City of Zagreb. Its position to the south of the Sesvete center, separated only by railway tracks, offers the advantage of establishing efficient rail and road connections to the city center of Zagreb.

The majority of the local population, gathered in the NGO "Green and Blue Sesvete," believes that this central space should be used for public facilities that are insufficient, in a district that records constant population growth. In the context of circular economy, several industrial buildings were analyzed for reuse as a community-driven hub, a music school, and

Figure 6.3 Green and Blue Sesvete Green Corridor plan.

other public functions. The discussion on reuse of the high silo buildings is still open, and scenarios of use are being explored. As industrial heritage, they have become a part of the local identity and Sesvete skyline.

Along with the Sesvete planning documents the Zagreb LL activities were informed and directed by the EU Green Deal policies as well as by the documents Sustainable Energy and Climate Action Plan of the City of Zagreb (Zurcher, 2020), Strategy for Adaptation to Climate Change in the Republic of Croatia for the period up to 2040 with a view to 2070 (2020), and Development program for Green Infrastructure in Urban Areas (2021) in Croatia.

Nature based solutions implementations in the Living Lab in Sesvete, Zagreb

In total, four out of the eight proGIreg nature-based solutions (NBS) have been implemented in the LL in Sesvete (Figure 6.1). Implementations focused on the following NBS: community-based urban gardens (NBS 3), aquaponics (NBS 4), and green roofs and walls (NBS 5), reusing derelict land for the new green corridors (NBS 6), and introducing low-carbon guidelines into new strategic urban planning documents (NBS 7).

Community-based urban gardens on post-industrial sites

The activities corresponding to this NBS are linked to the "City Gardens" project in Zagreb, established in 2013. The project was a result of initiatives by NGOs which approached the city government and together they

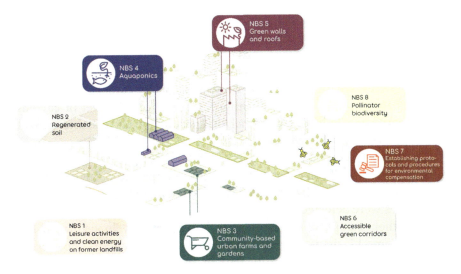

Figure 6.4 Implemented NBS in the Living Lab Zagreb in Sesvete.

Table 6.1 Overview of implemented NBS in Zagreb

Type of NBS	Description	Implementation
NBS 3: Community-based urban farms and gardens	NBS 3.1: Info point	Operating since the beginning of the project
	NBS 3.2: Therapeutic garden in Sesvete	Implemented in Spring 2021
	NBS 3.3: Modernized urban gardens	Implemented in Fall 2023
NBS 4: Aquaponics and NBS 5: Green walls and roofs	Modular urban farm	Implemented in 2021
NBS 6: Accessible green corridors	Pedestrian and cycling path across the LL area	Implemented in Fall 2023
NBS 7: Local environmental compensation processes	Local environmental compensation processes	Before and throughout the project duration guidelines for low-carbon development were incorporated into the strategic and planning documents.

84 Nature-Based Solutions for Urban Renewal in Post-Industrial Cities

co-created a successful model: gardens are implemented on city-owned land, and 50-square-meter plots are given to citizens free of charge for periods of two years. The "City Gardens" project is a successful example of sustainable land use in Zagreb, enabling citizens to produce food (vegetables and berries), herbs, and flowers for their own needs. City gardens, besides providing space for healthy food and improving the home budget of citizens, also offer the possibility of traditional food production and coexistence with nature, as well as social interaction.

There are two existing urban gardens in Sesvete. The one in Sljeme LL was upgraded with solar water distribution pumps and solar water purifying devices (the garden is watered with underground water in which certain spores were detected).

The other example of a community urban garden developed in proGIreg on the Sljeme LL site is the creation of a therapeutic garden. The users of the urban gardens have asked for part of the gardens to be accessible to people with disabilities, since there was considerable interest for therapeutic gardening. The new urban garden in Sesvete was designed to fit that need, providing accessible and serene space for gardening, relaxation, socializing, and inclusion, following the principles of sustainability. The therapists were included in the co-design stage to make sure that the garden suits the users' needs. The users are children from the Mali dom (Little Home) daycare center for children with multiple disabilities, for people with cerebral palsy, and people from the center for autism.

These users of the garden are involved in learning activities by planting herbs, vegetables, and flowers. The visitors to the garden observe nature with all senses, participating in various activities such as games and art classes with nature taking center stage. The Sesvete garden motivates families that have members with disabilities to engage in activities, increase interaction between those with and without disabilities, and integrate marginalized people into the local community. The users plant the vegetables, take care of them, nurture them, and in the end, harvest and consume them.

This therapeutic garden is an example of low-tech, bottom-up transformation of brownfield area into a green, sustainable, inclusive semi-public space that enhances spatial and social qualities of the neighborhood, and of the city as well.

The therapeutic garden in the LL consists of three zones:

1. area for user interaction (gazebo, grill, wooden platform/stage, and circular benches);
2. area for therapeutic gardening and education (elevated garden beds for growing herbs, hügelkultur beds, storage for tools and a trellis); and

3. sensory garden and sensory rest areas (elevated and classic beds with herbs, reflexology path, quiet rest areas, and interactive sculpture).

The activities are monitored and managed by the City Office for Economy, Environmental Sustainability and Strategic Planning. To enable users to regularly work in the calming environment of the garden, the city-owned daycare center for children with multiple disabilities "Mali dom" coordinates everyday use and maintenance of the garden. The therapeutic garden is, perhaps, the greatest success of the project.

NBS – aquaponics system and green roofs and walls

A modular urban farm is designed as an integral system that combines several nature-based and green technologies: green roofs, green walls, solar panels, and aquaponics. The implemented farm is a green technology hub in the Sljeme factory area, with twofold function: commercial and educational. The basic components of such a stand-alone system is a metal container unit of 36 square meters, with an example of green wall (two different

Figure 6.5 Therapeutic garden users, 2021.

Figure 6.6 Green wall on the Modular Urban Farm.

systems) and a green roof, powered by a solar panel: The aquaponic system is located in the container, with microclimate automation and control system and irrigation system.

The modular urban farm was used as a mini laboratory for food growing technologies, and as a case study for low-maintenance green walls and green roof.

NBS – reusing derelict land for new green corridors, bicycle lanes, and access to running water

Relocating the economic zone offers the opportunity to redefine the Sljeme LL area and redevelop the industrial area. In the center of the future mixed-use urban development that would connect the present Sesvete center and the existing Jelkovec housing development for 11,000 people, an 850-meter-long green corridor is proposed running along the planned street no. 6. The green corridor would offer extensive green areas, recreational facilities, a presentation of the old Roman road that once passed through and be an important part of the Sesvete green infrastructure. The existing therapeutic garden would be part of the future green corridor. It could also be a place to redirect the Vuger stream, whose present flow is being endangered by commercial development. This way the green corridor would also offer access to running water and other water features like ponds. This would be the focus of the new development in Sesvete.

As part of future street no. 6 and along the envisioned green corridor a cycling lane is planned that would connect with the existing cycling route on the north, built along the Vuger stream prior to the start of the proGIreg project, and later continue south to the River Sava joining the 121 km Greenway cycling route along the river.

During implementation, it became apparent that the planned road no. 6 will not be built in project timeline, so the partners have decided to plan another cycling lane along the road crossing the LL in the east-west direction.

An alternative corridor that connects the Slatinska street to the west with the zone of the Sesvete fair to the east was approved by the district council and implemented in the fall of 2023.

Figure 6.7 Tree planting activity nearby Info center, 2019.

88 Nature-Based Solutions for Urban Renewal in Post-Industrial Cities

Figure 6.8 Assisted natural regeneration activity, 2021.

Introducing low-carbon guidelines into new strategic documents

One of the project outcomes is the proposal of guidelines of low-carbon development planning, created in accordance with European green policies. These guidelines propose a new procedure for development and adoption of spatial plans through early participatory process involving citizens, and a turn toward low-carbon development, energy transition – decarbonization of the energy system, use of sustainable materials and circular economy, renewable energy sources, green infrastructure, sustainable water management, nature-based solutions, and renaturalization of urban centers. This new direction must consider assessment and management of risks and solutions for mitigation of climatic extremes effects and of natural disasters (floods, extreme precipitation, extreme droughts, etc.). The basic starting point is the transition from gray to green infrastructure, which is crucial for transition processes and builds on the basic starting point of spatial planning – the protection of natural resources.

The proposed guidelines for low-carbon development were incorporated into the Strategy for Adaptation and Mitigation of Climate Change in the Republic of Croatia for the period up to 2040 with a view to 2070 (Official Gazette 46/20) and the Sustainable Energy and Climate Action Plan of the

City of Zagreb (SGGZ 13/19) – SECAP. These documents set the direction for lower-level documents which will follow these principles and enable green development of the city and its surroundings.

In 2018, members of the proGIreg project took part in the Faculty of Architecture team to prepare for the Ministry of Physical Planning, Construction and State Assets a proposal for the Urban Green Infrastructure Development Program to support the green development in Croatia according to the EU Green Deal program. Sesvete (Sljeme LL) site was proposed among the suggested pilot projects. At the same time, the Faculty of Architecture also made the proposal for the Circular Management of Spaces and Buildings Development Program in Croatia, and the brownfield area of Sljeme in Sesvete was also suggested as a pilot.

Co-design activities in the Zagreb Living Lab with stakeholders and citizens

The proGIreg Info center was the first co-implementation in the LL in 2018, conceived as a complementary activity to the proGIreg project. The renovated building serves as a meeting point for the Sesvete community, venue for project meetings and workshops, and as a showroom to document LL processes regarding nature-based and sustainable solutions. The Info center is also home to the local chess club, thus merging different interest and age groups.

The Info center is run by the "Green and Blue Sesvete" NGO (ZIPS), which aims at promoting and raising the quality of life in Sesvete while supporting the urban transformation of the neighborhood. Aware of spatial shortcomings and problems (urban, traffic, planning, identity, recreation, brownfield sites, etc.) in its city district, the association started gathering influential and interested residents of Sesvete to actively take part in improving living conditions in Sesvete. Their motto is: "We live in Sesvete. Our families live in Sesvete. We must take care of Sesvete. Because Sesvete is our HOME, and only one place is called HOME."

From the start of the proGIreg co-design processes, consistent involvement of local stakeholders has proved useful for successful implementation while acting as a catalyst to detect underlying problems and risks. It also ensured that residents accept and embrace the co-created NBS. In this process, the local NGO Green and Blue Sesvete (ZIPS) played a crucial role based on its strong and active ties to the local community by engaging stakeholders in the decision-making process during several workshops and joint activities. Co-design and co-creation workshops provided a structured

90 Nature-Based Solutions for Urban Renewal in Post-Industrial Cities

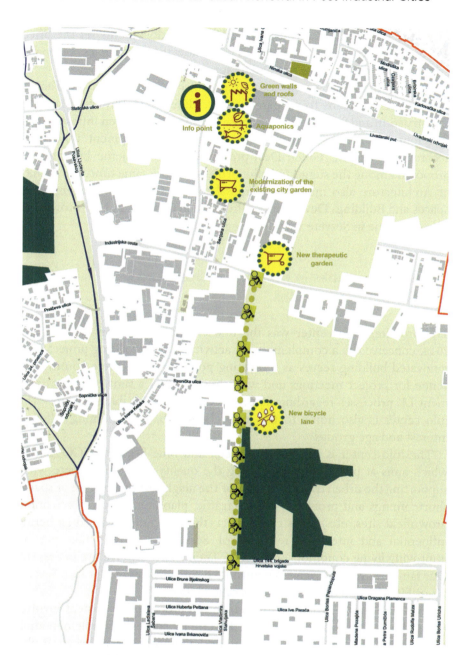

Figure 6.9 Workshop on farming, held at the therapeutic garden.

framework for involving citizens. The local partners expanded on that by inviting all relevant institutions to discuss the design of implemented NBS in the LL and additional activities.

Key objectives of NBS for inclusive urban regeneration on the former industry site of "Sljeme" include:

- Developing principles and architectural solutions for regeneration of the deprived and abandoned area.
- Developing nature-based proposals for public spaces that would promote urban activity, social cohesion, and reduce present insecurity in the area.
- Developing nature-based proposals for recreation areas and promoting a healthy and sustainable lifestyle.
- Define spaces of intergenerational interest.
- Developing urban plans for the zone that encourage public and private investment will result in a prosperous community.
- Developing makers, FabLab, and entrepreneurial programs for teenagers (HUB) to promote innovation and a new business culture.

Given the above objectives, stakeholders in the LL formulated the vision to utilize NBS to gently reclaim the area of the Sljeme former meat factory, using it as a platform for greater social inclusion.

A key measure of project success is how deeply it penetrates the hearts and minds of the local community, and whether it changes its behavior in the long run. ProGIreg inspired several activities that were not included in the initial project description. These actions have been managed locally by the NGO Green and Blue Sesvete (ZIPS), which has been instrumental in bringing the local population into the decision-making process of transforming the living conditions in the Sesvete district.

Based on the 2016 proposal, a pedestrian and bicycle path along the Vuger stream was realized in autumn 2019. During the COVID-19 pandemic, being the only place of possible recreation in the center of Sesvete, it showed its full potential by serving thousands of users and is continuing to do so in the post-pandemic period, accentuating the need for new public spaces and new cycling infrastructure. Furthermore, based on Zagreb LL commitments the ZIPS sent, for the entire Sesvete district, numerous other proposals for the pedestrian and bicycle infrastructure on a planning level or on a more detailed scale. They considered the potential of the rural character of certain parts of Sesvete (field and forest paths) and used the potential of preserved watercourses as a basis for development of a green infrastructure network.

Figure 6.10 Workshop on farming, held at the therapeutic garden.

In the period from 2019 to 2023, the proGIreg Zagreb team organized the planting of about 700 saplings of different types of trees and shrubs in Sesvete in cooperation with citizens and local organizations, thereby increasing the functional biodiversity of the area. Around 600 participants took part in the planting activities, mostly parents with children and members of various sports organizations. This activity was supported by numerous companies and organizations with financial donations, material, or work (Spajić, 2022; Zurcher, 2022). During 2021, the activity of cloning two centuries-old linden trees from the north of the city district started. Around 140 saplings will be grown and distributed to citizens and various institutions with the aim of preserving the gene pool of trees and to highlight the importance of genetic diversity.

The last example is related to the transformation of a construction material dump into a landscaped urban area. In cooperation with volunteers and local companies, about 500 cubic meters of discarded concrete and a large volume of other municipal and bulky waste were removed from an area of 12,000 square meters. Later, part of the invasive vegetation was removed, and the remaining vegetation was kept according to the model of assisted natural regeneration. Different flower species were seeded to serve the pollinators, and in fall of 2022, flowering trees and shrubs are planned to be planted on the western border of the area. This project created a green area for the benefit of citizens and nature, and is interconnected with other aforementioned project areas, hence creating *steppingstones* for future green infrastructure. ZIPS has organized many other waste removal activities from public space in the period from 2018 to 2022, and it has participated in numerous media appearances with the aim of preventing further pollution.

During the period of implementation of the proGIreg project, the NGO has had the opportunity to participate in three more EU-funded, and several other collaboration projects. Special mention should be made of the Erasmus+ projects – LE:Notre Landscape Forum 2019, ReGreen Zagreb 2021 and Learning Landscapes 2021, where the association and the partners from the city of Zagreb participated as a partner in the activities, in cooperation with the Faculties of Agriculture and Architecture as well as other European partners.

Since its establishment the Info point program has included lectures, workshops, exhibitions, and round tables. ZIPS is also very active on social media with a following of 22,000 citizens. The NBS topics have been recently extended and complemented with those of FabCity approach (EU horizon Centrinno project) and STEM (Stemerica project, STEM_Sesvete) which are being developed by ZIPS NGO, Fablab Croatia, and the Faculty of Architecture in Zagreb.

Therefore, the LL interventions have provided a public space oasis for the community, especially the vulnerable groups (users of the therapeutic garden). Local actions that accompany the project activities have also brought people into the area and provided them with a sense of belonging to the space.

Challenges and opportunities of the Zagreb Living Lab in Sesvete

Planned activities for Zagreb LL in the proGIreg project were significantly impacted by two mayor events: the COVID-19 pandemic and two earthquakes in 2020. The pandemic limited operations, as well as hindering citizen engagement under restricted conditions. Earthquakes caused the need to redefine where and how to develop green walls and roofs NBS. The initial building for this NBS had the financial support of the city for a reconstruction but was put on hold after the earthquake. The alternative plan was to develop a modular urban farm for the LL. But the global collapse of logistics chains worldwide resulted in shortage of modular containers due to their use in temporary accommodation for the earthquake-affected citizens, simultaneously increasing their prices. Most of the problems were connected to this NBS project and will be described within challenges.

Figure 6.11 Art exhibition in the Info center, 2020.

Figure 6.12 Learning Landscapes workshop, 2021.

Zagreb Living Lab **95**

Table 6.2 Overview of encountered types of barriers by NBS

Type of barrier	Description	NBS mostly impacted
Administrative/institutional	• Bureaucracy • Land ownership issues • Municipal utility availability in post-industrial areas	Green corridors Green walls and roofs Aquaponics
Technological	• Lack of expertise • Long-term maintenance	Aquaponics Green walls and roofs
Financial	• Additional budget required • Long-term maintenance	Aquaponics Green walls and roofs Green corridors
Social/cultural	• Lack of sense of community • Vandalism	Therapeutic garden

Challenges

Regulatory and legal compliance involves several key aspects. Local zoning regulations and the necessary permits presented a challenge, particularly the limited possibility of obtaining a permanent permit for setting up an urban farm due to the absence of an obligatory spatial plan. Ensuring safety standards was crucial, as the setup had to avoid violating environmental protection laws, especially regarding invasive species like tilapia.

The bureaucratic system posed challenges in connecting to the electrical power grid and obtaining permits, necessitating the creation of an off-grid system through solar energy and batteries. Infrastructure and resources were critical, with the need for a reliable power supply for lights, pumps, heaters, and other essential equipment. Additionally, the willingness of the local government for implementation was a concern.

Internal climate control and system design required addressing the challenge of temperature regulation due to the high albedo of the parking space and high sun radiation, necessitating adaptation to microclimatic conditions. The selection of appropriate vegetation and fish species had to consider conditions both inside and outside the container.

Accessibility posed a problem, as access for visitors and other users was difficult and limited. Data monitoring was inconsistent due to different methodologies among partners, and there was a challenge in sourcing available vegetation for cultivation, as well as other necessary materials.

Technical expertise was another significant factor. Monitoring and automation to optimize farm operations were hampered by site dislocation. The acceptability of available technology and knowledge for successful implementation was in question, and the use of new and, somewhat, unfamiliar technologies added to the complexity of the task.

Community impact included raising awareness and building support for urban farming initiatives, while also dealing with the threat of vandalism and theft. Accommodating the farm in a post-industrial area brought the challenge of engaging citizens, which required a high communication effort. Additional considerations were defining end users, market distribution, and decision-making in the management of the urban farm, especially after the end of the project. Financing posed a significant challenge, in the context of the project not being fully covered by the grant and concerns about the sustainability of the project after the end of the funding.

Opportunities

Opportunities included bringing the concept closer to citizens and sparking their interest in replicating technologies and production methods through popularization and awareness efforts. There was significant potential for involving the wider community and educating citizens, particularly through guided excursions for school children. The project also opened new possibilities for growing food and products, as well as allowed for the prototyping of products during its development. Establishing a short supply chain in production was another opportunity, supported by continuous involvement and support from active consortium members.

Case study – aquaponics and green walls/roof in Sesvete, Zagreb

In Zagreb LL, implementation of some NBS – namely, the aquaponics and green walls and roof, has met plenty of obstacles, which almost led to the NBS not being implemented at all.

The start of the proGIreg project coincided with the development of the design proposal for transforming the existing "Sljeme" factory administration building into a community-driven innovation hub. This hub was supposed to be a place that will promote digital technologies, makers' culture, creative initiatives, and local entrepreneurship. The design proposal, made by the Zagreb Faculty of Architecture, was planned to have green

Zagreb Living Lab   97

Figure 6.13 Installation of the modular urban farm.

walls, a photovoltaic installation, and an aquaponics system placed on the green roof. However, in the second year of proGIreg, both the COVID-19 pandemic and the earthquake that hit Zagreb in March 2020 made such a strain on the city's budget that it resulted in the hub project being put on hold. This delay in plans made it necessary to change the plans, as implementation would exceed the project's duration.

Therefore, an alternative plan for a modular urban farm was developed and implemented next to the therapeutic garden. The proposal for the mini urban farm combining NBS 4 (green walls and roofs) and NBS 5 (aquaponics) is based on the study "Investigation into sustainable implementation of nature-based solutions and the involvement of the local community."

The study analyzed three scenarios for implementation of green technologies: the implementation of green technologies within urban gardens, integration of green technologies into institutions with existing infrastructure such as schools or other public spaces, and creation of a new and integral system that would integrate green roof and walls technologies with aquaponics. The third scenario was evaluated as the most favorable for achieving the proGIreg goals.

The modular urban farm was developed by "Vesela motika" (Happy hoe) company, which is experienced in innovative urban farming and indoor solutions. The modular farm combines green walls and roofs technologies with aquaponics powered by a solar panel. In the process, several roof and wall plant species were tested for the Zagreb climate to be later recommended for wider use (Lugović, 2020).

Figure 6.14 Workshop on farming, held at the therapeutic garden.

Prior to the installation of the aquaponics system, detailed analysis resulted in a realistic business model that could make the system appealing to the local people. The installed modular container urban farm provides a showcase for the application and practical features of these technologies.

The modular farm operated during the project timeline, but with limited capacity as the electricity fixture wasn't available on site. Therefore, the aquaponics never really took off, but the hydroponic indoor farm was operating and growing young trees and plants such as basil. Several workshops took place in the modular urban farm, giving the interested public and local farmers an opportunity to learn about the small scale farming and urban gardening, informed both by traditional and biodynamic techniques and by innovative and technologically advanced solutions.

After the project ended, the modular urban farm is still functioning and its legacy has spread beyond the proGIreg project. Another project, UP2030 (https://up2030-he.eu/), funded by Horizon Europe and expanding on NBS, is using and upscaling the solutions tested before, making sure that the research done through the project stays relevant and thrives on.

Zagreb Living Lab – from NBS to an urban regeneration process

Despite growing recognition for the potential of NBS, operationalizing them into policy and urban planning as well as facilitating their implementation requires extra work, as they usually cost more than the traditional gray infrastructure-based construction. The LL in Sesvete is a showcase of some of the NBS, giving the community and wider public a preview of the expected green and sustainable future.

Modernized urban gardens are easier to use and demonstrate a new green technology; therapeutic garden is more of a social innovation, bringing the disadvantaged people into the community and some serenity and gardening therapy to them. Modular urban farm is a living and breathing research facility run by people who are happy to share their experience, and the cycling path is another probe penetrating a formerly barricaded land. All NBS experience is gathered and shared in the Info center that has been the meeting point of all the project-related activities since the project's beginning.

The post-industrial zone will have to be transformed in the following years, and this has to be powered by the local initiatives and shaped by the needs of the community. The co-design process has shown the potential of public participation in the planning of the environment, and we hope that the people in Sesvete will use the tools developed in the project. The undertaken NBS and related interventions in the Zagreb LL in Sljeme have inspired and encouraged local citizens to continue with their own initiatives and activities.

The Road 6 (the originally planned cycling path) will be developed in the future as a traffic corridor that includes the cycling track which will be built in accordance with the green and sustainable principles, and the cycling network will have more potential to expand and develop, providing sustainable and environmentally friendly mobility to the area. All the implemented NBS will continue to function and provide potential and inspiration for replication – especially the therapeutic garden, which has attracted the attention of city planners on an international level.

With the conversion of the abandoned industrial facilities in Sesvete (Sljeme and Badel), a wider urban regeneration of Sesvete can begin, providing spaces for needed public facilities and spaces, and creating a contemporary identity of the district. Empty silos and interesting industrial buildings need to be reconstructed and reused. It is certainly important to turn industrial plants into mixed use areas (housing, commercial activity, hospitality, work, recreation, etc.).

In the future, Sesvete must base its development and urban plan on the highest level of environmental protection and sustainable development. Traffic solutions, housing, economic zones, agriculture, energy production, waste management centers, water purifiers, and recreation zones must illustrate community awareness that this space is intended for the grandchildren

100 Nature-Based Solutions for Urban Renewal in Post-Industrial Cities

of the current inhabitants. For the whole of Zagreb, Sesvete can become a living laboratory for exploring and promoting nature-based and advanced technology solutions for the present ecological challenges. This approach should generate the educational, cultural, and development interest of the wider community and be a model for Croatia.

References

Jošić M., Baletić B. (2016). *Polazišta projekta Zelene i plave Sesvete* (Green and Blue Sesvete – Project Starting Points), Faculty of Architecture University of Zagreb.

Lugović S. (2020). *Istraživanje održivosti implementacije rješenja temeljenih na prirodi i uključivanje lokalne zajednice* (Sustainable Implementation Research of Nature Based Solutions and Local Community Participation), Vesela motika.

Spajić M. (2022). *Green Infrastructure and Planning Procedures – Experience of Creating an Unofficial Network of Green Spaces as an NGO*, pp. 76–80, Acta Horticulturae et Regiotecturae.

Zurcher N. (2022). *The Global Experience in Green Participatory Place-Making: A Worldwide Sampling of Urban and Community Forest Connectivity*, pp. 127–133, Future City 16.

(2021). *Program razvoja zelene infrastructure u urbanim područjima* (Development Program for Green Infrastructure in Urban Areas), Official Gazette 147/21,

(2020). *Strategy for Adaptation to Climate Change in the Republic of Croatia for the Period up to 2040 with a View to 2070*, Official Gazette 46/20.

NBS spatial analysis processes

7

The role of spatial analysis for Front-Runner Cities and Follower Cities

Sabina Reichert, Oana Emilia Budău and Codruț Papina

Spatial analysis process

Building a common framework

The purpose of a spatial analysis is to understand and explore the entanglement of the spatial positioning of objects and phenomena and their characteristics (Audric, de Bellefon and Durieux, 2018), being an important instrument for the study of spatial phenomena and the relationships between them. ProGIreg cities developed a multi-level, multi-dimension analysis, on the basis of existing administrative spatial data, for the purpose of highlighting the current level of spatial development and pre-conditions for implementing nature-based solutions (NBS) in Living Labs (LL) (Front-Runner City [FRC]) and for developing urban regeneration plans (Follower City [FC]). Spatial analysis is considered a relevant process, in which cities are mapping the factors to which NBS

DOI: 10.4324/9781003474869-7

This chapter has been made available under a CC-BY license.

relate to: socio-economic aspects, environmental conditions of the neighbourhood and climate factors, urban planning priorities and strategic frameworks.

The analysis of spatial data, namely the observations with a known value and location, represents a complex and rich source of information offering important insights into the context in which NBS will be deployed. This process relies on already-collected data and allows for an understanding of pre-existing local spatial dependencies, correlations and trends which can ultimately be factored into the ex-post evaluation of tested NBS.

Without additional resource-intensive data collection, a useful and synthetic understanding of local sub-municipal, urban and metropolitan contexts can be achieved via the collection and analysis of existing indicator data, corroborated with information derived from the existing multi-level plan and policy landscape. Unpacking the spatial manifestation of social, economic, environmental, political and administrative factors that could potentially enable or hinder the implementation of NBS can reveal practical and realistic "entry points" when designing interventions that contribute to nature-led urban regeneration.

In this chapter, we look at the spatial analysis process carried out with the 4 FRC and 4 FC in proGIreg, its results, and the way in which collaboration and stakeholder engagement have effectively supported comparability between the cities and a more robust understanding of baseline conditions.

Regardless of the planning stage of the NBS interventions, the spatial analysis is an absolutely necessary process to ensure: (i) the efficient insertion and adaptation of the NBS; (ii) the creation of a synergy or links at the local level of the NBS, from the point of view of the spatial configuration and the urban structure with the neighbourhoods, but also from the point of view of coherence with other urban factors; and (iii) ensuring the construction of a viable and valuable strategy, which takes into account all the "driving forces" for a "green" transformation in line with the needs of the communities and the complex local landscape.

As a result, one of the first outputs of the project was this analysis component, carried out for both FRC and FC alike (even if the end purpose of the analysis is different). The process was aimed at developing a common and comparable spatial framework based on existing indicator data, corroborated with information derived from the existing multi-level plan and policy landscape in each city.

FRC and FC roles

The role of FRC in proGIreg is to adapt and test the NBS locally, in clearly defined post-industrial neighbourhoods with related socio-economic challenges (Huckarde, Mirafiori, Sesvete). The FC Cascais, Cluj-Napoca, Piraeus and Zenica target future implementation of these NBS through their integration in the local urban planning framework. The utility of the spatial analysis in FRC was to articulate the NBS experimentation carried out in LL in the local urban structure for the neighbourhood and city level (the location of NBS in FC was decided before hand, based on a preliminary assessment of needs and requirements). In this way, the NBS benefit assessment performed after their implementation is contextualized and well-grounded to the local state-of-the-art. In contrast to FRC, where NBS location was already defined, the purpose of FC analysis was to create an evidence base identifying key entry points and pre-conditions for new solutions, further co-developed in regeneration strategy, presented by the FC in the form of an urban plan – mapping clear actions in clear defined locations, representing the outcomes of intensive co-design process.

Methodology of the spatial analysis component in proGIreg

The analysis approach is multi-level, because it defines two analysis scales used to collect, process, analyse and assess administrative spatial data and data on plans, policies and stakeholders. It is also multi-dimensional, assisting cities to develop a basic spatial development baseline for four key assessment domains: (1) socio-cultural inclusiveness, (2) human health and wellbeing, (3) ecological and environmental restoration and lastly (4) economy and labour market.

These categories represent the factors to which NBS either directly generate an impact (such as ecosystem services and environment) or are relevant factors to which NBS planning or implementation has to take into consideration (climate conditions, access to services, local communities, etc.). An important aspect of NBS, which makes the analysis process a must, is the "solution" component. In order to implement "solutions" (that are nature-oriented or nature-inclusive), one must understand the "problem". And the problem, to which the NBS can respond to and mitigate it, is composed of one or cumulus of factors from these domains.

Any empirical spatial analysis is concerned with a finite bounded region. The spatial analysis is sensitive to both characteristics of the zoning system

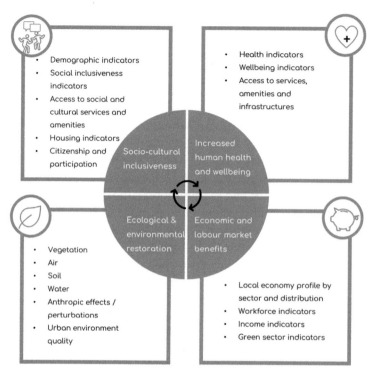

Figure 7.1 ProGIreg key assessment domains.
Source: ICLEI

used to collect the data, as well as the scale at which data is reported (Fotheringham and Wong, 1991). Identifying this boundary and determining the proper unit of analysis is dependent on the scope of the spatial analysis, the issue of data availability, complexity and time.

The limit of the analysis area represents a complex issue, subject to an ample body of research which has underlined the boundary problem of spatial analyses: namely, while geographical study areas are bounded, spatial processes are not. The consequences of this misalignment may lead to inaccurate results, or improper understanding of causes and effects (Ripley, 1979; Fotheringham and Rogerson, 1993). The so-called *edge effects* can be reduced or eliminated by enlarging the analysis area and creating a buffer zone in which data is also examined in terms of their effect on the area of interest.

When deploying NBS implementation or planning NBS (regeneration) strategies, the analysis process has to be deployed at multiple levels, from macro to local or detail scale: city level and neighbourhood level, and, additionally, intervention site-level. The importance of the results of the analysis and the effort made for the various scales of analysis vary according to the purpose of the project: the amplitude of the NBS, the target users, the location of the intervention and overall the objective of the initiative. The general recommendation is that, no matter the case, one should go through all the scales of analysis, to ensure that the NBS fits in the chosen context, and not only that – a carefully thought out multi-scalar analysis can add value to the NBS by connecting with various components/actors at the urban level. Also, a key differentiating point in the effort put in the analysis at different scales is the level of "experience" of cities/local groups have with greening initiatives. For cities which are only just beginning the journey of planning greener neighbourhoods, a more in-depth analysis of driving factors for generating a sustainable transformation might be required, but as practice shows, one of the most important aspects is capacity-building and raising awareness (for making NBS the new standard).

For data availability purposes, the administrative boundary of the metropolitan area and the city was considered for the first level, while the second one was analysed at sub-municipal (district) level, wherever data was available (Turin, Zagreb, Dortmund and Cascais). Additionally, in the process of elaborating urban regeneration strategies (FC-focused process), analysis at the site level was needed in order to establish key aspects of NBS design for future implementation.

FRC were able to better articulate the planned NBS in relation to the local context, with the main outcome of the experimentation being a proper estimation of the generated impact and the knowledge on how to implement NBS more easily and faster. FC analyses were the first step for a better understanding the local challenges and NBS suitability assessment to the chosen neighbourhood context, ranging from neighbourhood-level (Cascais) to river corridors across the whole city (Cluj-Napoca) as well as dispersed locations in specific city districts (Piraeus, Zenica). FCs were in the need of performing additional analyses, more qualitative and oriented on communities. Thus, the process of elaborating FC urban regeneration plans resulted in several iterations of the study limit, a back and forth between conclusions/outcomes of the co-design process and updating/detailing some components of the analysis.

Analysis methodology

The methodology of the spatial analysis followed an incremental approach, divided into three phases, closely interrelated – setting the analysis the framework, collecting valuable data, assessment. Added value of the elaborated methodology is that it ensures a high level of replicability for cities in similar contexts – with the objectives of implementing NBS. The constructed methodology is ensuring an efficient assessment of the impact/evolution of the interventions, in the context of the neighbourhoods.

Phase 1: Setting the analysis framework. Spatial analysis methodology had the aim of guiding cities in performing meaningful analyses and data collection, in relation to: (1) the green infrastructure (GI) and environmental conditions, (2) post-industrial neighbourhood regeneration and (3) quality of life. Setting the analysis framework means that one must delineate what is relevant for deploying the planned actions, or delineate what to analyse in detail in order to construct feasible and valuable decisions for future investments. In short, proGIreg had a specific focus on post-industrial neighbourhoods – because the main objective and research direction of the project is how NBS can be adapted in these types of contexts. Thus, the scope of the analysis has to be in line with overarching objectives of the city, and that strongly relates to the political factors: the public administration (or relevant public bodies) has to recognize development (or regeneration) priorities, not only from economical point of view, but also from social and environmental. Depending on the purpose of the research, phase 1 can be framed as: (1) co-design effort of partners in order to commonly agree on the methodology for collecting data – a valuable recommendation for audience that activate in research projects; (2) preliminary assessment of relevant factors to be analysed, depending on the scope of the project; and (3) always focusing on well-being (to which ecosystemic services provided by GI and NBS are strong contributors or enablers).

Phase 2: Collecting valuable data. The analysis framework is being enriched with information, with the aim of providing a solid baseline for the initial situation, highlighting: (1) potential drivers for implementation and (2) potential challenges. An important recommendation to cities, planners and researchers is to make an effort to correlate hard data (spatially-bound or not) with: (i) community-oriented information (needs and priorities, intangible heritage understanding [local identity, landscape heritage], behavioural patterns and also openness towards NBS); and (ii) empirical observations: planners and researchers must not be disconnected from the reality of the space itself – even though NBS experimented in proGIreg are usually small-scale interventions, a good understanding of the spatial configuration of the neighbourhood, local landscape and ambiance,

Phase 3: Assessment. Critical conclusions and qualitative analyses were performed, in the form of SWOT analyses, in order to synthesize the important contextual elements, better understand driving factors and potential barriers and risks and also map the key aspects to take into consideration when transitioning to co-design activities. Translated to other analysis processes, phase 3 is about drawing up conclusions and constructing a synthetic understanding of challenges, opportunities and drivers.

Aspects to consider for NBS transformation at city level/URA level

The proGIreg project provides a flexible, easy-to-adapt set of eight NBS solutions, each with special features, requiring different resources, generating different spatial impacts and overall functioning in a different way. Thus, the spatial analysis can differ according to the scope of the cities in the process: scenario 1: analysing the urban environment for articulating a specific NBS (e.g., productive gardens or green corridors), based on already acknowledgement of challenges; scenario 2: analysing the urban environment for assessing the local challenges and constructing NBS aimed to mitigate it. Nevertheless, the set of spatial data and indicators collected must be useful for decision-making in both the design phase and in the strategic planning phase.

Nevertheless, it is recommended that NBS are considered in conjunction with existing strategic and regulatory framework and the relevant stakeholders when engaging in spatial analysis process. Depending on the challenge identified, NBS relate differently to the provisions of existing plans, policies and regulations, to key stakeholders and to resources, since the NBS require different levels of financial resources, depending on the NBS typology, scale of interventions and target users.

Developing a spatial analysis to support NBS design and deployment relies on the three aforementioned phases, which generate an approximation of the initial situation and conditions. However, the spatial analysis is not a baseline in the strictest sense defined by literature, because it offers a current "snapshot" of development, planning and governance framework relevant for NBS implementation, dependent on how experienced a

Table 7.1 Research questions guiding the spatial analysis

Research questions	Method	Data sources, collection and analysis
Is there an enabling regional and/or local strategic, programmatic, regulatory and normative framework that can support NBS and urban regeneration plan development?	Qualitative survey on the existing plan, policy and regulatory frameworks at the regional and local level, screening for the degree of support (implicit, explicit) for key GI and NBS concepts	Strategic, programmatic, regulatory and normative documents: city survey, desk review, consultations
How do the NBS correlate with the territory and stakeholders planned to be included in the co-design processes?	Inventory of key stakeholder groups in the city, by NBS typology of interest	Stakeholder identification information (type, institution) from the city and secondary sources (other key stakeholders)
What is the current socio-cultural, human health and well-being, ecological, environmental, economy and labour market level of development in the city and/or neighbourhood? How does that translate spatially, at the two analysis scales?	Quantitative survey of context data (state and process) for the four key assessment domains. Development of SWOT analyses, with inclusion of qualitative fact-based assessments. Spatial analysis of synthetic SWOT maps to illustrate the four components at city and/or neighbourhood level.	Collection of statistical and spatial data from existing sources (municipal databases, national/regional census, etc.). Stakeholder input for desk analysis and SWOT (text, illustration).
What is the overall context of spatial development, from the point of view of the four key assessment domains? What would be important focus points in implementation?	Review of sectoral findings. Development of conclusions	Interpretation and final conclusions. Optional validation with stakeholders.

NBS spatial analysis processes 109

Figure 7.2 Methodological steps of the spatial analysis.
Source: Dimitriu, Elisei (2020)

city is in experimenting with GI and NBS previously. The policy context, resource mapping and existing stakeholder analysis consequently need to be an integral part of the spatial analysis, due to the importance of cross-referencing what can be implemented with where, in which context and with whom:

Regarding data availability, the lack of statistical spatial data can hinder the creation of a sufficiently robust profile for one or several of the key assessment domains. Generically, data required for the analysis can be categorized as:

1. Spatial data: Available geodata, based on the Infrastructure for Spatial Information in Europe (INSPIRE) Directive (2007) Data Themes, targeting spatial data which can be used in environmental studies, planning framework and policy design. These can be maps, either raster or vector – computer data files (GIS, dwg, etc.) or remotely sensed data such as satellite imagery or orthophoto plans.

2. Statistical data sets pertaining to the spatial analysis scale of the city and the LL/urban regeneration area, collected as tables, graphs and charts.

The main data sources for the spatial analysis are municipal databases, data from service providers, external stakeholders (e.g., business registers, NGOs), regional and national data from statistics institutes or census data, existing local and supra-local documentations and grounding studies (e.g., previous air quality studies for SUMPs) or other databases at European level such as EUROSTAT, ESA Copernicus, Europe's soil database and ECMWF.

One of the main challenges in spatial analyses represents data variability and lack of reliable data. This can be addressed by cities through different strategies:

- **Exclude initially considered indicators** from the list if they are unavailable, provided that the database still contains a coherent group of primary indicators across assessment domains;
- **Define proxy data** or replace initial requested data with similar alternatives, available in the respective city; and
- **Include the indicators if they are critical**, marking their values as not available for the spatial analysis but performing their collection throughout implementation and monitoring.

Lastly, beyond data availability, the usefulness of spatial analyses resides in the capacity of the cities themselves to interpret data, spatially represent it and derive conclusions.

Structural components of the spatial analysis

Basic data

It is useful for cities at first to provide a descriptive part of their development context and general introduction, a high-level overview of the context in which regeneration plans and NBS investments will be carried out. This will assist the identification of levels of analysis, as well as the NBS to be implemented or those identified for potential implementation at this stage, allowing for the formulation of the first intervention hypotheses. More in-depth and focused analyses will afterwards be performed during the co-creation process of elaborating the urban regeneration plan.

Plan and policy framework

NBS have a wide-ranging impact across key assessment domains and interact in practice with many different other solutions and policies at urban level. Policy coherence is thus required between the implementation or embedding of NBS in the urban system and the existing strategies, plans, policies and regulations in force in the cities. Furthermore, a successful implementation of the eight NBS requires a strong integration with the cities' existing governance practices, institutional and regulatory frameworks.

Cities performed a comprehensive review of multi-level existing normative and strategic plans, as well as other city investments or actions in potential synergy with planned NBS are important, looking into urban development, GI, urban regeneration, participation and social inclusion. Findings from the experience of the eight proGIreg cities' analyses indicate supportive planning frameworks for all the FRC and coherence of the selected NBS. Local NBS adaptation can be a first step in having large-scale initiatives – Dortmund and Turin analysis concluded that synergies can be created with wider region or city-level strategies.

In FC, previous experience with NBS is limited; hence, proGIreg offers the opportunity to embed the solutions in new plans and policies. There are concrete possibilities to contribute to the development of spatial planning documents in FC Metropolitan area of FC Cluj (County/Metropolitan Plan) and FC Zenica (Green City Action Plan), under development.

Stakeholder identification

Broad cooperation in NBS planning, implementation and innovation allows a shift towards a systemic, open and user (beneficiary) centric policy. The quadruple helix approach of proGIreg is recommended as a way to involve four key stakeholder groups: civil society (NGOs and individual citizens), academia (universities and research institutions), governmental institutions (local governments and other public authorities) and the private sector. The quadruple helix enables cities to foster and sustain NBS, as well as the public acceptance and uptake, economic viability and sustainability. In this part of the analysis, the stakeholders should be identified and described, generally making sure of their representation of all primary interest groups. Further stakeholder mapping workshops can fill in the gaps and go beyond the "usual suspects" of stakeholder participation.

Spatial data indicators

Quantitative data collection relied on a cooperative indicator framework developed in the project, supporting the definition of a common spatial analysis framework that can be adapted by other cities, beyond the scope of proGIreg, to facilitate analysis, comparison and decision-making.

Spatial analysis of statistical indicators (understood as the representation of statistical data for a specified time, place or any other relevant characteristic) considers the fact that data often exhibits properties of spatial dependency and explicitly includes these considerations in the formulation of conclusions. The indicators should be selected to transmit information about the state, or the state evolution (variation) of a phenomenon which cannot be measured directly and that can be impacted by the deployment of NBS. They should allow the perception of differences (territorial disparities, improvements or developments related to a desired change or in a certain context), related to NBS implementation or the formulation of urban regeneration plans. Wide-scale deployment of innovative NBS can have a potential effect on an ample set of urban parameters ranging from the quality of the environment to the development or substantial growth of a local green job sector. It is thus necessary to survey available indicators from administrative databases that pertain to the four key assessment domains (socio-cultural inclusiveness, human health and wellbeing, ecological and environmental restoration, economic labour market benefit) and to select a robust database comprising at minimum of two to three state and pressure indicators for each sub-domain.

The indicator collection process can show heterogeneous availability of indicators, with, for example, very limited data in what concerns information on the health of the population and the environment quality (air, water and soil). In this case, cities should consider whether this data can be derived from other proxy indicators or whether it is useful to explore other avenues such as population surveys or regional data collection.

Swot analyses

Based on the spatial analysis of statistical indicators, as well as data available from other sources such as plans and policies, SWOT (Strengths, Weaknesses, Opportunities and Threats) tabular analyses represent a very useful tool to synthetize findings at both local intervention level as well as urban/metropolitan level. SWOT offers the possibility to condense different elements of an urban audit into a comprehensive picture, and to analyse alternative scenarios of urban and territorial development, and it is a good choice for cities due to its versatility and ability to represent in an organized way the influence played by multiple factors on different decision contexts (Comino and Ferretti, 2016). From the methodological point of view, the SWOT analysis is structured into:

1. Internal environment analysis: endogenous factors (variables that are part of the system and that can be directly modified in a desired way through NBS implementation or planning); and
2. External environment analysis: exogenous factors (variables external to the system, but which can influence it, and that can either be valorized as opportunities or mitigated by NBS investment, if their impact is negative on the internal environment).

It is useful to spatialize findings of the textual analysis in thematic SWOT maps, which can be layered to reveal concentrations of multi-domain spatial and socio-economic issues or strong points.

- SWOT analysis for NBS planning, implementation and monitoring must incorporate a robust spatial or territorial component. Information or conclusions from data analysis that cannot be directly spatially represented should be linked to various spatially determined factors or conditions within the city or neighbourhood.
- Identified weaknesses in the SWOT analysis can be transformed into opportunities through the adoption of the right approach and strategic thinking. By definition, a weakness is an inherent characteristic of the internal environment, whereas an opportunity arises from the external environment. Consequently, to convert a weakness into an opportunity, one must harness external resources, foster innovation and improvement, discern socio-economic trends and facilitate collaboration and partnerships.
- SWOT analysis can allow for more targeted interventions that align with the unique characteristics of the neighbourhood. Thus, performing SWOT analysis before NBS implementation or during the NBS planning process, one can better articulate the solution to socio-economic requirements, environmental conditions and local landscape.

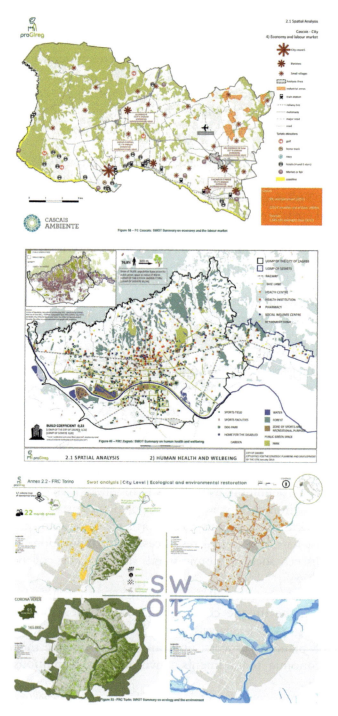

Figure 7.3 SWOT analysis extract: methodological steps of the spatial analysis.
Source: Dimitriu, Elisei (2020)

Analysis outcomes and lessons learnt

In short, the analysis process outcomes are in the form of:

Front-Runner Cities outcomes

Table 7.2 Front-Runner Cities outcomes

	Main findings
Dortmund	• High amounts of green spaces per capita • Complex renaturation projects of post-industrial brownfields and the Emscher river • Local economy successfully transitioned from industrial activities, 80% of population being employed in tertiary sector
Turin	• Important green areas with good distribution at city level, connected to the wider belt Corona Verde, which gives the city a high degree of ecosystemic services and various options for NBS adaptation • Existence of poor-quality urban areas: social deprivation, including difficult access to housing, high unemployment, low school attendance rates and low levels of vocational education • In the context of deprived neighbourhood, community gardens represent a valuable asset.
Ningbo	• Poor connectivity of GI elements • Unsatisfactory air quality • Moon Lake Park – a valuable facility for the residents, providing important ecosystemic services, but in need of renaturing and restoring the ecologic properties of the lake (thus performing bioremediation of the contaminated waters is a priority)
Zagreb	• Abundance of green and natural spaces in and around the city, natural, historical and cultural monuments, and well-preserved built heritage • Increasing poverty and social stratification – urban gardens are very valuable as means for sustainable and affordable food source • Sesvete neighbourhood lacks a proper integration in urban frameworks – industrial areas are still disconnected and represent both a valuable asset for future development, and a factor that creates fragmentation in the urban tissue.

Follower Cities outcomes

When the methodology was provided at the beginning of the project, FC identification of regeneration area and, consequently, set of NBS, was still at an incipient level. Therefore, the first conclusions of the spatial analysis have been reviewed, adapted and, in some case, re-elaborated at the beginning of the replication process in FC.

Conclusions

Depending on the scope of various actors engaged in urban analysis for NBS implementation or NBS planning, the level of detail and effort can vary a lot. Nevertheless, the challenges are: (1) establishing the focus and scope of the analysis; (2) collect valuable data and correlate it; and (3) construct the final assessment by keeping an objective view on the matter.

The contexts of the cities differ widely, but between FRC, the approach has been to test NBS in post-industrial neighbourhoods (Huckarde, Mirafiori, Sesvete), having particular socio-economic challenges, for which NBS can make a strong case towards supporting renewal and redevelopment.

For FC, this spatial analysis offered an opportunity to establish a first area-based approach to the urban regeneration plans. Their options have produced a diverse approach, with FC Cascais identifying a peripheral neighbourhood of the city, similar to FC Piraeus, who provisionally delineated two such areas in the Western and Eastern sides of the city. FC Zenica and Cluj both opted to select areas on the courses of rivers (Bosna and Somes, respectively). While the first one has narrowed the potential urban regeneration plan area in the heart of the city, the latter has the ambition of including the entire main development axis of the city (along the Someş river and railway) within the activities of proGIreg. This heterogeneity represented an opportunity to test the NBS in very diverse settings, at diverse scales, and to validate them in multiple replicable contexts.

The plan and policy framework analysis represented a critical step in realistically assessing options for NBS deployment, due to the trans-sectoral character of NBS effects and their multi-scalar nature. Analysing gaps or opportunities in the multi-level governance support (policies, plans and regulations) before implementation mitigates risks in how these solutions would ultimately be financed and maintained. In FRC, horizontal and vertical integration is ensured in all cities, with Dortmund leveraging the most on the existence of an overarching GI and NBS development concept. FC

NBS spatial analysis processes **117**

Table 7.3 Follower Cities Outcomes

	Main findings
Cascais	• Uncontrolled urban expansion has led to problems including weak pedestrian accessibility, a lack of urban green areas, fragmented urban tissue resulting from the expansion, river pollution and restricted river accessibility for local residents. • The study area is divided in two neighbourhoods that have poor connections due to the highway that acts as a barrier. There is the opportunity to make the river corridor accessible (that runs underneath the river) and connect these two areas. • River pollution and limited river accessibility for residents. The rehabilitation of the river corridor will unlock a series of plots for comprehensive adaptation of NBS as a driver for the regeneration of local landscape and socio-economic development. • The area consists of series of valuable sites that are considered "locked" plots, being privately owned. There is the need for a better implementation of the local land policy Cascais Land Bank, which allows authorities to exchange terrain with private owners for the subject of urban agriculture.
Cluj	• City transformation towards innovation, business, youth and culture at regional and national levels, marked the city's motto "Greener Cluj", driving important investments and initiatives in the past years. • The FC Cluj-Napoca URA spans from east to west along natural and industrial axes. There is relevant potential to embrace green urban transformation through the utilization of existing elements and conversion of neglected and underutilized land. • Initial identification of three challenging areas for FC Cluj URA: industrial and rail axis, blue-green axis of the Someş River and secondary corridors: Nadăş and Someşul Mic, utilization of neglected/underutilized green areas present in social housing areas and collective housing neighbourhoods.
	• The city and URA is surrounded by valuable natural landscape, to which the future NBS investments have to relate and create a coherent green infrastructure system from territorial level towards city level. • Spatial analysis guided FC Cluj to choose NBS6 as a key intervention for regeneration of industrial and railway areas, and NBS3 as suitable option for revitalizing the local landscape of social and collective housing areas.

(Continued)

118 Nature-Based Solutions for Urban Renewal in Post-Industrial Cities

Table 7.3 (Continued)

	Main findings
Piraeus	• FC Piraeus is among Europe's densest urban environments (38% housing units <30 m²), lacks significant green infrastructure (0.83 m²/person) requiring urgent greening for climate adaptation. • City analysis maps guided FC Piraeus to target NBS3 in public institutions (schools, kindergartens), envisioning transformed school yards into community-managed vegetable and plant gardens, also with pollinator capability. • Lack of connection between communities and nature. Piraeus seeks to foster a sense of community with nature by involving students in implementing and managing school-based community gardens, supported by teachers.

present different contexts, and while some of them have previous experience with NBS (e.g., urban community gardening in Cascais), others are newcomers to the topic, with general and implicit, rather than explicit support for NBS. Nevertheless, in contexts where there is no experience in working with NBS, the development of a well-framed and participatory urban regeneration plan can provide an opportunity to explore, together with the quadruple helix, not just how to find the right solutions for current urban needs, within the existing regulatory framework, but also how to identify financial support – either in partnership with the stakeholders involved or more commonly in cohesion areas, through mainstream funding.

Regarding stakeholder engagement, while the initial identification is useful to the extent that it frames the general scope of interventions (e.g., involving schools in Torino), the process is very much an iterative one throughout each step of co-design and implementation. The most changes in stakeholder groups have been observed in FC, as their groups have expanded and became more specific to the particular solutions to be included in the LL implementation regeneration plans.

Regarding data collection, one of the biggest challenges starting from the beginning was ensuring a critical mass of (existing, already collected) indicators on which to base the spatial analysis. This approach has produced at least one to two valid indicators per sub-domain for each city, but has challenged the possibility of ensuring comparability. While in the context of the project research, the latter was relevant, it would not be a challenge impacting stand-alone spatial analyses for NBS deployment in other contexts. However, even data for FRC has been particularly limited, especially in what concerns health and environmental quality assessments, which underlines the importance of collecting disaggregated health and environmental data at urban and metropolitan level, vital especially in the post-COVID-19 context.

Spatial analysis of all eight cities, disregarding the difference between what constitutes an FRC and an FC, was an important step that facilitated a smoother replication/transfer process of good practices from FRC to FC. The FC could assess the similarities between the identified challenges and determine what type of approach could ameliorate them. Given the focused scope of FRC (implementing and monitoring NBS at the local level) and the broader approach of FC (developing a regeneration strategy), it can be stated that the challenges and contexts of the FC were more diverse – nevertheless.

The cities exhibit considerable differences, yet commonalities persist among them. High levels of air pollution, partly due to traffic, are evident in Zenica, Cluj-Napoca and Piraeus. Challenges such as very high population density, private ownership of brownfields, overcrowding and a lack of urban connections or relationships with green areas are characteristic issues shared by Cluj and Zenica. Deficient pedestrian and bicycle accessibility represent a problem identified across all four cities. Cascais faces challenges similar to those of Turin/Mirafiori and Huckarde, with a potential urban regeneration area marked by a lower-income social situation, low education, discrimination and illegal soil occupation with diverse functions. Cluj, in contrast to all three FRCs, has an economic challenge: property costs are high and remarkably so, with the city leading in upward rent and land/construction costs at the national level. Furthermore, while Zenica is experiencing significant depopulation, Cluj is growing, albeit at a slower pace akin to Dortmund.

These are all crucial factors that must be further considered in the development of urban planning processes in the follower cities. Their diversity ultimately represents an advantage for proGIreg, enabling the testing and exploration of the integration of NBS in various settings.

Finally, the SWOT analysis confirmed cross-cutting issues characteristic of post-industrial and socially deprived areas such depopulation, economic stagnation, social segregation and disconnection, lack of public services and low urban fabric permeability, coupled with a higher incidence of several diseases (e.g., Mirafiori) and pollution (all FRCs). While the scope of the spatial analysis was to provide data-supported evidence for the main urban phenomena in these areas prior to deploying participatory processes, it was also an instrument for building the co-design process itself, as the desk analysis lent itself to a process of stakeholder validation and discussion, a critically useful tool for the first workshops in FC aimed at problem analysis, in which the participants used the desk analysis findings and completed or challenged data-derived conclusions. We can conclude that one of the most valuable steps of any type of urban and landscape analysis is the participative component that heavily influences the way the project or the strategy is taking shape.

References

Audric, S., de Bellefon, M. P., & Durieux, E. (2018). Handbook of Spatial Analysis: Theory and Application with R. INSEE Méthodes N°131. French National Institute of Statistics and Economic Studies. Available at: INSEE website

Comino, E., & Ferretti, V. (2016). Evaluating Sustainability In Uncertain Contexts: A Method To Support Decision Processes In Territorial Transformation Projects. *Environmental Impact Assessment Review*, 56, 87–98. https://doi.org/10.1016/j.eiar.2015.09.007

Fotheringham, A. S., & Wong, D. W. (1991). The modifiable areal unit problem in multivariate statistical analysis. *Environment and Planning A*, 23(7), 1025–1044. https://doi.org/10.1068/a231025

Fotheringham, A. S., & Rogerson, P. A. (1993). *GIS and Spatial Analytical Problems*. London: Taylor & Francis. ISBN: 9781857280991

Ripley, B. D. (1979). *Spatial Statistics*. New York: Wiley. ISBN: 9780471722516

Co-design NBS with post-industrial communities

8

Margot Olbertz, Bettina Wilk, Israa Mahmoud, Emanuela Saporito and Ina Säumel

Introduction

Nature-based solutions (NBS) address complex urban spatial, social, ecological, and economic challenges simultaneously and can deliver multiple benefits such as alleviating air pollution, climate mitigation and adaptation, improving biodiversity, and social inclusion. There is wide agreement that NBS benefit from engaging communities in the design, planning, implementation, and maintenance of NBS. A diverse set of actors is required to collaboratively design NBS that adequately respond to and reflect different needs and are fit to the local context. To identify where and how NBS can tackle existing spatial and social challenges tailored to local needs, citizens' ideas, experiences, and knowledge are the starting point for stimulating and challenging urban developments. Key to success is integrating insights of the local environment to develop NBS that are widely valued and accepted. Living Labs (LL) can serve as real-life testing grounds for innovative NBS throughout co-creation processes. These processes span different phases from planning, co-design, exploration/co-implementation to the evaluation of NBS benefits (see Figure 8.1). Given multi-level stakeholder constellations in co-creating NBS, diverging interests can potentially be reconciled to achieve socially just and inclusive solutions and outcomes in each phase.

DOI: 10.4324/9781003474869-8

This chapter has been made available under a CC-BY license.

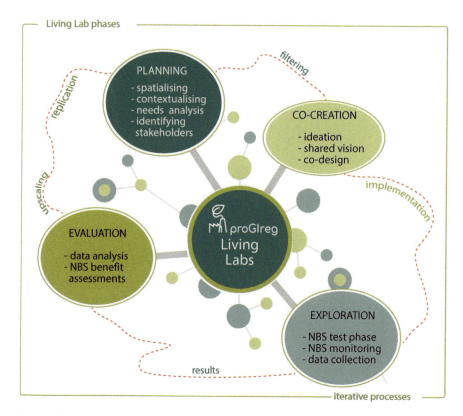

Figure 8.1 Steps of setting up an urban Living Lab (RWTH).

Co-creation of NBS represents the core of generating valuable context-, needs-based and community-driven solutions that are sustainable at different scales in the urban context. Co-creation processes, and co-design in particular, aim at achieving mutually valued outcomes and benefits, and joint ownership of implemented NBS. This active participatory process systematically involves relevant stakeholders from diverse backgrounds of local governments, academia, private-sector actors, and local communities based on the quad helix approach (Figure 8.2), as early as possible in the process, to the very end. However, the level of engagement intensity and the type of stakeholder may vary by co-creation phases. For instance, the local community may be less engaged in the planning or building permit application stage for technically, administratively complex NBS, but more in co-implementing and notably co-maintaining NBS. When engaging local stakeholders as equal co-creators during co-design, co-implementation, and co-maintenance process phases can result in unusual stakeholder constellations. Co-creation processes, and co-design in particular, aim at

Figure 8.2 Quadruple helix model (ICLEI).

achieving joint ownership of NBS implemented and a good fit between NBS and local needs.

However, such collaborative, trans-disciplinary processes, and procedures for joint NBS design and implementation are still novel to traditional planning policy and practice. Frantzeskaki et al. (2022) point out that for such an integrated urban planning approach, "a paradigm shift in the approach to design is necessary", challenging traditional governance arrangements and entrenched institutional landscapes. The experimental and inclusive objectives of NBS require new participatory structures and mechanisms, skillsets, organizational resources, relationships for exchange and collaboration, and ultimately policy frameworks. Reconciling and mediating the iterative co-design/co-creation process often benefits from professional facilitators to help build trust and a shared understanding of its benefits (Hölscher et al., 2020).

Thus, experimenting and testing NBS in real life and safe spaces in the form of LL has become a common method in NBS pilot design, implementation, and potentially highlights options for scaling up NBS in other urban contexts. LL exhibits a specific type of experimental (urban) governance in which local stakeholders jointly develop, test and evaluate new technologies, products and services to produce innovations solutions to urban challenges (Voytenko et al., 2016). Post-industrial proGIreg partner cities experimented with and tested participatory and trans-disciplinary planning processes to establish different types of NBS in such urban laboratories in collaboration with local authorities, communities, businesses, start-ups, and researchers. Their aim is to generate urban sustainability transitions by strengthening democracy and social justice through user involvement in urban regeneration areas or neighbourhoods where social, economic, and technological ideas and concepts are developed.

This chapter is structured as follows: Firstly, outlining the concept and methodological approach of co-design and co-creation in the proGIreg project, delineating it from traditional forms of stakeholder engagement. Secondly, highlighting identified governance arrangements for co-creation and zooming into the different roles and responsibilities public and private actors can take in co-creation regarding design, implementation, and management of NBS. Thirdly, case studies explore practical implementations, lessons learnt, and game changers of locally adapted co-design processes in proGIreg LLs, and sister projects will be presented while exploring the most important barriers across NBS design, implementation, and maintenance identified in proGIreg. Experiences and strategies for overcoming barriers shed light on underlying governance arrangements. Lastly, case studies provide concrete examples of co-creation processes for different NBS and governance models.

Concept and methodological approach to co-design and co-creation in proGIreg

Demarcating co-creation from stakeholder engagement

Stakeholder engagement is understood as the

> the process of building relationships with people and putting those relationships to work to accomplish shared goals, i.e., involving those

who are at the heart of the change we wish to see. Achievement of excellence in such engagement practices can be through a high quality of work in conducting research, building partnerships, and co-constructing and mobilising knowledge for achieving sound impact.

(Tandon et al., 2016)

Co-creation goes further than stakeholder engagement in traditional participatory processes in urban developments. Following Voorberg et al. (2015), proGIreg defines co-creation as

the active engagement of actors who hold different types of knowledge and resources with the aim to generate collaborative [and mutually valued] outcomes [and benefits...] Outcomes can vary and can include vision narratives, new understandings of problems and opportunities, hybrids of solutions.

Key differences between stakeholder engagement and co-creation are:

i) degree and intensity of stakeholder engagement in the process or producing knowledge, and designing and implementing solutions; and
ii) maturity of strategy underpinning stakeholder engagement during the life cycle of co-design, co-implementation, and co-maintenance of NBS.

In proGIreg, co-creation is a strategic approach to engaging different stakeholders during all phases of establishing NBS (from co-design, co-implementation to co-maintenance/co-management). It is based on systematically involving the local community and other stakeholders as early as possible in the process to the very end to ensure continuous community engagement. Still, municipalities often shy away from engaging the civil society early, notably in case of unresolved issues in the beginning. But in the light of greater planning transparency, stakeholders may gain a better understanding of urban planning challenges and perceive changes to their environment as less disruptive. Organizing stakeholder engagement around questions of intensity of involvement and the impact of different stakeholder groups on processes and results (i.e. NBS) is exactly where co-creation conceptually differs from more traditional forms of stakeholder engagement and public participation (Voorberg et al., 2015). Fundamentally, the civil society's role shifts from being observed subjects to a source of creation, empowering citizens to influence innovation processes and its results.

In practice, stakeholder engagement is a spectrum rather than a concrete process. It can range from information, consultation, involvement,

126 Nature-Based Solutions for Urban Renewal in Post-Industrial Cities

Table 8.1 Spectrum of government and non-government roles in different governance arrangements in the nature-based solutions (NBS) Living Labs

Government actor role	Leading		Involvement	Partnership	Enabling	None/ regulatory
Form of Non-government actor participation	Information	Consultation	Involvement	Partnership	Empowerment	
Non-government actor role	Provide information about nature based solutions (NBS) projects as part of decision making process		Some involvement in planning, management, maintenance of NBS	Shared roles & responsabilities around NBS planning and management	Leasing or purchasing of public resources to implement NBS	Management agreement, leasing or purchase of privatly own resources to implement NBS
Governance model	Government actor led model		Co-management	Co-governance	Non-government actor led model	
Modes adapted from Driessen et al. (2012): Government (G); Private Sector (P); Civil Society (CS)	Government primarily responsible for NBS activities with some stakeholder engagements (i.e. participatory planning or budgeting)		Government main responsible for NBS. Other support management	Many different actors involved: equal role of network partners	Private sector or community organisations are lead actors with the government taking a supporting role (i.e. facilitating resources)	

Source adapted by authors: Wilk et al. (2021).

partnership to empowerment (Table 8.1). These gradients differ by the extent of power and influence public and private stakeholders have on decision-making processes and the development of the final solution. The further to the right, the more balanced the power distribution between private stakeholders and public authorities becomes. Co-creation has a more limited operating space than stakeholder engagement, but instead fosters civil society involvement, partnership, and empowerment.

The quadruple helix model

The quadruple helix model is based on four equal stakeholder groups with diverse roles, allowing for collaborative, interactive, and transdisciplinary engagement of citizens to generate mutual outcomes.

In proGIreg, the starting point for co-creating NBS is a multi-stakeholder arrangement in the LLs, established by the quadruple helix partnership of local partners (see Figure 8.3). It integrates academia and research institutions, local government representatives (municipality), the private sector and industry (SME implementing the NBS and entrepreneurs), citizens, and NGOs as civil society representatives. Given co-design is a dynamic process, additional stakeholders (i.e., other municipal departments,

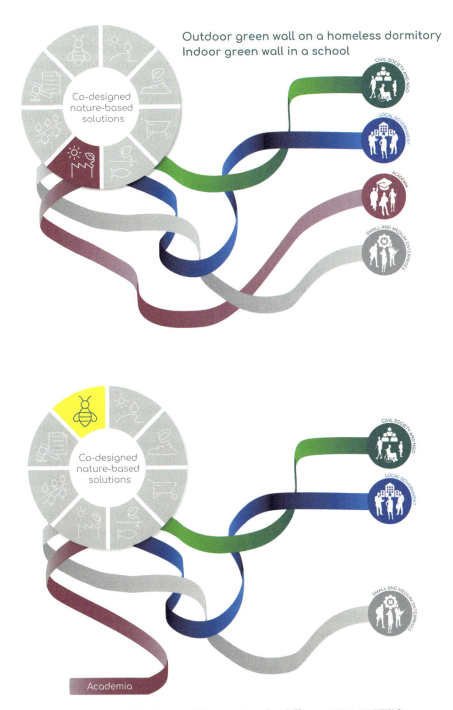

Figure 8.3 Quadruple helix model in practice for different NBS (RWTH).

128 Nature-Based Solutions for Urban Renewal in Post-Industrial Cities

foundations, and cooperatives active in the LL area, school administrations, public institutions, real estate companies, landowners, property managers, and so on) may engage to varying degrees throughout the process phases.

Based on a literature review of different co-creation approaches (grey literature and H2020 NBS projects resources, status 2020), proGIreg identified and validated six co-design principles with the local actors in the LL (Wilk, 2020). These principles aim to facilitate and guide participatory processes in co-creation-oriented planning and decision-making, are flexible in application, and can be adapted to any specific local context:

Openness, inclusiveness, and diversity: Co-design processes should be organized by inviting and addressing different types of knowledge, perspectives, and needs to design fit for purpose, context-specific solutions. Making activities open and accessible can improve the quality of stakeholder engagement and allow for a fairer distribution of benefits emerging from the NBS.

Shared goals and vision: Stakeholders are bound to have different expectations and interests of managing the implementation of NBS. Discussing and seeking to align differing expectations in a joint vision is a pivotal step in the co-design process to finding common ground and a shared understanding of the project's aims, goals, and needs among stakeholders engaged.

Transparency: Transparency from the start of the co-design process helps to avoid disappointment, builds trust, enhances ownership of NBS, and ensures good relationships between stakeholders. Stakeholders should be aware of the procedures, rules, and framework conditions, such as timeline, objective, and rules of conduct, and should know what is expected from them and what level of influence in the co-creation of NBS each one has.

Long-term thinking: Long-term thinking in NBS management processes and planning is pertinent. Good initial design and planning can tackle arising barriers to ensure long-term success of NBS. This includes monitoring strategies and instruments, long-term lease contracts of NBS, as well as taking measures for long-term maintenance of the infrastructure and services through citizen co-ownership arrangements.

Experimentalism and reflectiveness: NBS require a great deal of experimentation and learning which occur in iterative cycles. Environments in which NBS are co-created should allow participants to create and test new solutions and services in real-life contexts. Such environments demand acceptance of trial and error, as well as learning from failures as part of the process.

Flexibility: Flexibility in the co-design process and its activities is key to ensuring adaptability to changing needs and priorities of stakeholders and

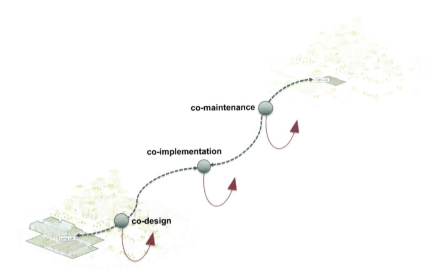

Figure 8.4 Co-creation process phases (RWTH).

adjusting strategies if needed. Flexibility should be safeguarded in topical issues, agreed objectives, plans, and activities, interaction processes (e.g., working modes), and structures of collaboration (e.g., ways of exchange).

Co-creation processes: These are broadly split into three main phases (see Figure 8.4) whose boundaries tend to be blurred since they are non-linear and often iterative. For instance, unforeseen hurdles such as regulatory issues (i.e., no visitors allowed in green houses, animal protection rights that prevent the use of fish in aquaponics, etc.) might prevent co-implementation or co-management from moving forward and thus force stakeholders to go back to the drawing board, i.e., the co-design phase.

1. **Co-design**, as the initial stage of co-creation, defines problems and tackles local challenges by collaboratively designing NBS interventions, e.g., developing ideas for the envisioned future outcomes, NBS benefits and strategies at suitable locations, iterative mapping of relevant stakeholders to co-create context specific NBS. *(Note: General planning may precede co-design.)*
2. **Co-implementation** entails putting the co-designed solution into action including physical constructions by involving local stakeholders, e.g., creating pollinator-friendly spaces and implementing green roofs.
3. **Co-maintenance/co-management** of NBS should ideally be planned in the co-design stage to ensure that citizens and stakeholders are motivated and willing to co-maintain and co-manage (depending on the governance model) an NBS in the long term. Possibly it requires extending the stakeholder network.

Governance models underpinning co-creation

Following the quadruple helix model, co-creation raises questions about roles and responsibilities of public and private actors regarding design, implementation, and management of NBS. A key objective of applying co-creation approaches in the proGIreg LL is facilitating a paradigm shift in tackling urban regeneration and green infrastructure development and experiment different governance at local level. Traditionally, urban planning and urban development processes are governed by public actors, namely local governments and local authorities in a top-down approach. In contrast, the co-design approach applied in proGIreg's NBS activities aims at increasing citizen engagement in planning and management processes on different urban scales, while acknowledging that public goods, such as nature, can be managed more effectively when both public and private actors are involved (Zingraff-Hamet et al., 2020). This shift implies changing roles and distributing responsibilities of governmental actors and authorities from leading to enabling (see Figure 8.5). However, co-design processes certainly increase decision-making time but shorten implementation time, ensuring greater success in the long run.

Through its underlying principles of inclusion and diversity, co-creation also implies a certain operating space for the governance of NBS involving local governments and authorities (G), the private sector (P), civil society (SC), and academia (Uni). Building on the work in proGIreg, its sister projects CLEVER Cities and EdiCitNet (Wilk et al., 2021) classified three governance models underpinning co-creation (Table 8.2), excluding the government-led actor model as a traditional model of urban planning:

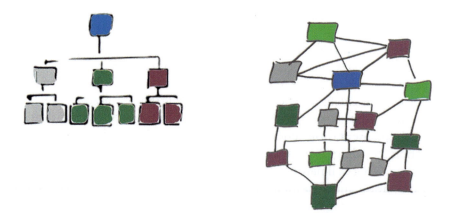

Figure 8.5 Co-design approach shifting roles from top down (left) to distributing responsibilities (right) (RWTH).

Table 8.2 Governance models for NBS underpinning co-creation with examples from proGIreg's Urban Living Labs including beforementioned case studies and additional samples

Type of governance model	Example
In **co-management**, a public actor (G) carries the main responsibility for NBS implementation while its role can vary between a coordinating, consultative, and/or supportive one; a private (P), e.g., business, or civil society actor (CS) supports the planning, management, and/or maintenance of NBS, often in a predetermined role (contractual agreements, public–private partnership, etc.), often on public space.	• Zagreb's Therapy Garden is the result of public–private collaboration. The Strategic Planning Department of the City of Zagreb is the overall coordinator, the Office of Agriculture and Forestry oversees its management. • A local NGO representing the user group is key intermediary and is entrusted with the operation of the therapy garden, including provision of materials and supplies. • Besides the partnership with the local NGO Green-Blue-Sesvete (ZIPS) with vast experience in citizen engagement in the district, several local NGOs working with targeted marginalized groups act as strategic partners and bridging organizations.
Co-governance represents a form of interactive governance with many different actors involved in participatory public-private governing arrangements. It is characterized by largely equal actor roles in design, implementation and/or management of NBS. Partnerships can be formalized or non-formalized.	• "Farfalle in ToUr" in Turin represents an interactive governance type where several public (universities) and private stakeholders (Local Health Company; users of Mental Health Centres) are involved in the design and implementation of pollinator-friendly green spaces and perform largely equal roles. • The department of Life Science and Systems Biology of the University of Turin has the lead role in the design and implementation, together with the users of the Mental Health Centres as main implementers. Formal partners are Mental Health Centres and educators of the social cooperatives with whom all activities are coordinated. • City of Turin is supporting the initiative.

(Continued)

132 Nature-Based Solutions for Urban Renewal in Post-Industrial Cities

Table 8.2 (Continued)

Type of governance model	Example
Non-government actor–led model represents a form of self-governance and is characterized by a bottom-up approach and participatory private-private governing arrangements. Roles are largely equal and power relations balanced across all actors while rules and procedures are often informal; the private sector (P) or civil society (CS) are lead actors and the public actor (G) takes a supporting/responsive role.	• A multi-stakeholder and transdisciplinary approach in planning and implementing an urban food forest in the Dortmund LL in a bottom-up approach. • The collaboration between a local NGO, a church community and a university faculty activated civil society engagement without regulatory framework and administrative procedures. • Given NBS activities took place on church-owned land, verbal agreements between church council and NGO sufficed to co-design, co-implement, and co-manage the food forest. • Church members, other residents, and the boy-and-girl scouts of the LL in Dortmund-Huckarde use and maintain the NBS with shared responsibilities.

Source adapted by authors: Wilk et al. (2021).

Co-designing NBS in proGIreg's Living Labs

Engaging multiple stakeholders in co-creating NBS harnesses the collective intelligence and local knowledge of a community, using the full potential of people available. This in turn creates a sense of citizen ownership, local identity, and new local solutions (Voorberg et al., 2015). This sub-chapter provides experiences and insights of locally adapted co-design processes in the proGIreg LLs, driven by the following key questions: (a) which stakeholder groups have a valid interest to be engaged, (b) what can the different stakeholder groups bring to the process and what impact can they have, and (c) how to engage and varying levels of intensity of involvement during co-creative processes by phase and type of NBS? It will also touch on barriers encountered during co-design and possible solutions or even game changers.

Identifying and interacting with target stakeholder groups

Central to co-designing NBS is engaging key target stakeholder groups that have a valid interest in transforming their living environment and will benefit from using nature for urban regeneration or may affect decision- and policy-making at a larger scale. In preparation of co-designing/co-creating locally adapted NBS within an LL, it is important to gain an understanding of the local context on various scales and parameters to adopt tailor-made strategies for integrating NBS. Conducting a thorough spatial analysis in each LL highlights urban regeneration needs and challenges of specific areas, indicating how to respond with targeted measures and pointing to potential partners and beneficiaries of NBS. A key task of co-design is identifying, analysing, and mapping context-specific stakeholder networks to facilitate the stakeholder and citizen engagement throughout all phases. Such mapping can help discover stakeholder clusters and unveil power structures, which allows for assigning roles and responsibilities along the following criteria:

(i) information and resources stakeholders bring to the process;
(ii) influence stakeholder capacity to affect the NBS and decision-making;
(iii) interest each stakeholder has in the NBS; and
(iv) impact the NBS might have on them.

Initially, proGIreg cities created a local group of core stakeholders that have a strong impact on developing NBS and can facilitate diverse stakeholder dialogues, then mapped other relevant local stakeholders and multipliers during co-creating NBS and defined main target groups (see Figure 8.6).

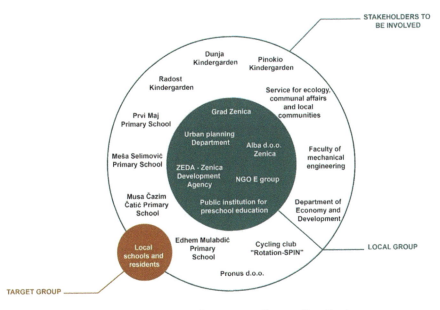

Figure 8.6 Initial stakeholder constellation in Follower City Zenica (URBASOFIA).

Figure 8.6 depicts the local authorities as enablers in co-creating a vision to integrate NBS in urban regeneration processes by allocating public resources, providing strategic leadership, coordinating, and not least facilitating the networking across administrative units, while closely collaborating with representatives of the local civil society, universities, and other local stakeholders such as private-sector actors according to the quadruple helix approach.

Stakeholder engagement needs to be planned systematically by mapping all relevant stakeholders according to the envisaged type and intensity of involvement. The Participation Planner (developed by EU H2020 research project CLEVER Cities) offers a structured approach and proved to be a useful tool to map existing and planned stakeholder engagement in the context of LL, and the envisaged involvement (from passive "recipients" to active co-creators). It highlights five levels of potential engagement (inform, consult, involve, partner, and empower) and different methods of engaging with the four quad helix stakeholder groups within these levels, focusing the co-design process as early as possible.

Given each involved stakeholder enters the co-design process with its own agenda of possible outcomes and approaches, jointly developing a shared vision and concept for the LL/Urban Regeneration Area acts as a catalyst to leverage original ideas conveying the added value for all. A long-term vision can be drawn, using imagery or storytelling, or formulated as a slogan or mission statement (Figure 8.6). Frequently revisiting and adjusting ensures responding to changing needs and stakeholder priorities. When vision and goals are developed bottom-up, residents are more likely to take ownership of NBS, ultimately ensuring the sustainability of NBS.

Based on a differentiated analysis of stakeholders and key users, tailored engagement formats can then be developed for each location, NBS, and group of stakeholders to benefit from specific knowledge and networks for implementing and gaining long-term citizen ownership. Mapping stakeholders per NBS also contributes to establishing realistic expectations of their involvement. At the same time, the co-design process needs to seek synergy potential between NBS for best possible impact. Creating transparency from the start of the co-design process is imperative to avoid disappointment, building trust, generating citizen buy-in of NBS, and ensuring perceived fairness of the process. It raises awareness of procedures, regulations, and frameworks, such as timeline, objectives, rules of conduct, and scope of influence, that enables flexible adjustments.

From the onset, proGIreg aimed at involving vulnerable, marginalized groups that are often difficult to reach in typical participatory planning processes. Key driver proved the involvement of NGOs or local groups who

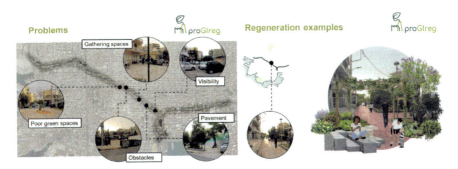

Figure 8.7 Co-design results from Follower City Piraeus, Greece (URBASOFIA).

mediate communication between vulnerable groups and key stakeholders such as local authorities. Reaching out to vulnerable stakeholders and responding appropriately to the special needs of homeless, refugees, or people with mental and physical disabilities has created valuable and targeted NBS interventions in proGIreg's LLs (see case study Zagreb Therapeutic Garden).

Intensity of stakeholder engagement by co-creation phase, type of NBS, and government model

The variety of NBS makes a universal approach to co-creation almost impossible (see Mahmoud & Morello, 2021). Therefore, approaching each NBS type and its operation at different scales separately has proved paramount. For instance, high-impact large-scale NBS infrastructure limits the engagement of the wider public, while small-scale NBS participatory processes may waste resources (Mahmoud & Morello, 2021).

The degree of stakeholder involvement may differ largely between NBS, limiting co-creation in design, implementation, and maintenance. The level of intensity in the co-creation process is influenced by three key factors:

- the governance model employed (mostly determined in the early co-design phase), see Bradley et al. (2022);
- the level of technical expertise and planning complexity required for certain types of NBS – may be relevant in polluted and/or contaminated post-industrial areas; and
- the scale of NBS interventions, e.g., multi-stakeholder engagement in very small-scale interventions might waste resources, but high-impact large-scale NBS infrastructure must include the wider public (Mahmoud & Morello, 2021).

Depending on the type of NBS requiring a certain expertise or knowledge at a particular stage of co-creation, the involvement of each stakeholder may change during the co-creation process: a stakeholder may be very active in co-designing the NBS according to local needs, but steps back during physical construction of the NBS to re-emerge in co-managing and co-maintaining the implemented NBS. Taking the NBS tested in the proGIreg LLs, generally NBS such as urban gardening or improving pollinator biodiversity are well-suited to involving communities throughout all co-creation phases.

The Orti Generali urban garden in Turin and the Food Forest in Dortmund are prime examples of involving mainly civil society stakeholders in active citizen engagement from co-designing in the conceptual phase to co-implementing until the handover and maintenance of the NBS (Figure 8.8). Often these NBS are motivated by non-governmental bottom-up initiatives in which local authorities play a secondary role in setting up and running the NBS. However, the Therapeutic Garden in Zagreb's LL represents an interesting example of merging a grassroots and top-down approach (see case study).

NBS requiring more technical expertise in the planning process and implementation along their life cycle tend to show lower citizen involvement, in particular in the early phases of co-design, e.g., building green roofs or green walls, aquaponics systems, creating new urban soil, or developing green corridors (Figure 8.9). Building permission applications, landowner issues, and knowledge of special technologies limit citizen engagement. Experiences from the LL in Turin demonstrate that once the green wall at a homeless shelter in Turin was coordinated and built by the municipality, citizens can be trained to co-maintain established NBS interventions. It has since stirred high interest in caring for the plants among regular guests of the shelter and from residents.

Barriers to co-creation and possible game changers

Inevitably, co-creating NBS in the proGIreg LLs encounters communication difficulties as well as administrative, technical, social, and financial barriers. Transparent co-design processes should be geared towards identifying and flagging technological and non-technological barriers of NBS at an early stage so that mitigation strategies can be devised.

ProGIreg's review and analysis of non-technological barriers across the phases of co-design, co-implementation, and co-maintenance/co-management of NBS[1,2,3] (Latinos, 2020) offer two important insights into barriers for co-creation:

Co-design NBS   137

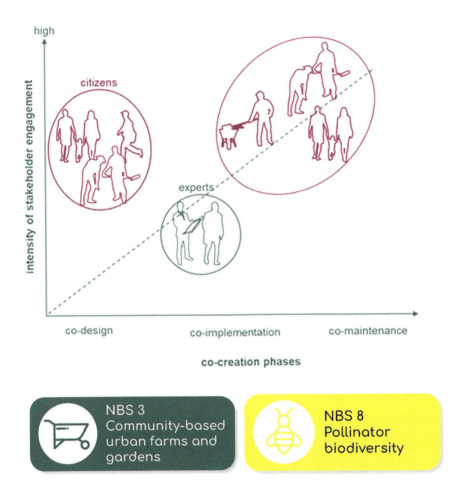

Figure 8.8 Intensity of stakeholder engagement by type of NBS and non-government actor-led model (RWTH).

I) It identifies the occurrence of barriers according to NBS co-creation phase, e.g., different barriers occur specifically in the co-design phase.
II) It enlists not only institutional barriers but also social and cultural barriers that either directly (e.g., lack of social acceptance of NBS) or indirectly affect co-design processes and determine important framework conditions (e.g., technical expertise required, land use requirements, spatial planning procedures).

Cross-referencing the different barrier categories with co-creation phases shows variations in occurrence and prominence across the phases (Figure 8.10).

138 Nature-Based Solutions for Urban Renewal in Post-Industrial Cities

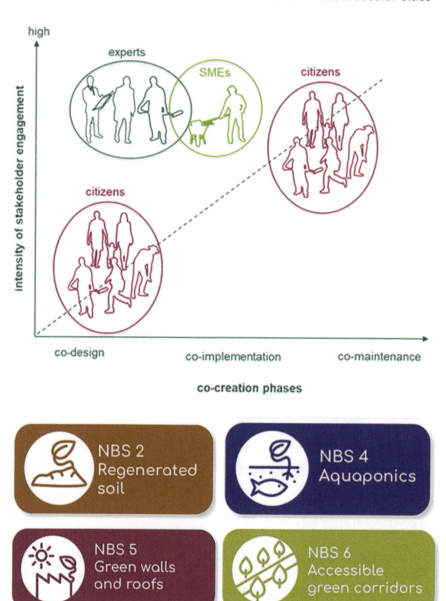

Figure 8.9 Government actor–led model (e.g., NBS green corridors, green roofs/walls, new urban soil, aquaponics) (RWTH).

Co-design NBS   139

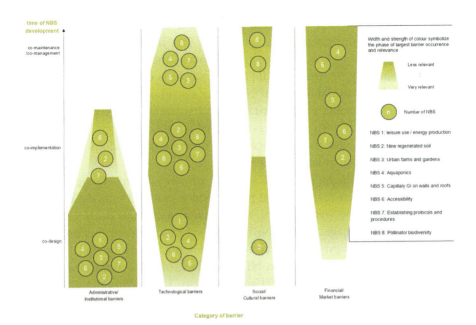

Figure 8.10 Key barriers by NBS co-creation phases and type of NBS (SWUAS, 2021).

Institutional barriers occur most prominently during the co-design phase of all tested NBS in proGIreg, as opposed to co-implementation. The most frequently mentioned and most decisive institutional barriers are extensive and time-consuming bureaucracy in building application processes, along with lack of previous experience and established procedures, mixed land ownership (public/private ownership), and lack of communication between municipal departments (sectoral silos) (Latinos, 2020). Extensive bureaucracy in the first stages of establishing NBS can delay the start of the co-design process through unforeseen regulatory barriers, such as building codes, safety measures, or fragmented land ownership in the desired location of NBS. Considering that NBS are rather novel solutions, the lack of previous experience and established procedures as a reference poses a pertinent barrier and complicates processes, limiting co-creation with citizens and stakeholders at an early stage. This also makes transforming technical knowledge and tested practices into easy-to-upscale NBS at city level more difficult.

Furthermore, sectoral silos and entrenched administrative structures can hamper co-design processes of NBS. For instance, often different departments govern and manage green spaces (one department is responsible for planning processes while maintenance is dealt with in another). The multi-dimensional nature of NBS may also elicit differing, potentially conflicting economic, environmental, and societal interests that cannot be or are not dealt with appropriately during NBS co-design.

Technological barriers occur across all three phases but dominate during co-implementation, as is to be expected.

Social and cultural barriers appear mostly in the co-design and co-maintenance/co-management phase, which can partly be explained by the nature of respective barriers. Key social and cultural barriers include lack of citizen engagement in NBS including ineffective inclusion of citizens in local decision-making and engagement already in the co-design phase, limited public interest in NBS, and lack of NBS mainstreaming in traditional plans and practices (Latinos, 2020). Limited public interest in NBS and social acceptance is often due to a lack of understanding and thus citizens' awareness of the multiple NBS benefits (but also city councillors). As a result, environmental benefits of NBS are often underestimated by the public. Targeted public campaigns that emphasize the multiple benefits of NBS for a city or district are an effective tool in raising awareness and empowering citizens to put small-scale natural solutions and/or biodiversity enhancing measures in place. However, mounting pressure on greening cities to tackle the multiple urban challenges, municipalities must be more transparent of all layers of planning to sensitize the civil society for the opportunities but also inherent difficulties of urban regeneration with NBS.

While mentioned barriers point to persistent structural, policy-related, and administrative bottlenecks for co-creation, several cities have successfully developed mechanisms to overcome them. So-called game changers include transformative shifts in framework conditions from hindering to enabling. For instance, changes in municipal planning and decision-making procedures through embedding co-creation and co-governance processes in urban planning practice (Mahmoud & Morello, 2021), as well as changes in policies and regulations. The integration of mandatory greening measures or co-creation criteria in tenders for new housing development to safeguard the implementation of green roofs and biodiversity-enhancing spaces on buildings are some examples of creating more flexible structures for co-designing NBS with the civil society.

Case study: Turin – governance models to empower local communities

The city of Turin implemented 17 of the 8 proGIreg NBS, each with different approaches to co-creation and civic engagement processes. NBS featuring innovative technologies such as "New Soil", "Aquaponics", and "Green walls" engaged relevant stakeholders (private companies, p.a., universities) in a mainly top-down government-led approach.

Other NBS tested (community gardens, green roof, green corridors, improving pollinator diversity) showed community engagement from co-design to co-management. The possibility to configure a non-governmental co-creation process of NBS was possible for some cases. Local NGOs such as the Community Foundation of South Mirafiori or OrtiAlti having professional competencies on NBS for community participation acted as mediators in facilitating civic participation at the LL level while being reliable partners for the municipality. Experienced citizens contributed to the successful co-creation process for low-tech NBS such as community gardens, which were implemented in different contexts including schools, public squares, and social housing courtyards.

On the administrative level, the local administrative tool "Regulation on collaboration between citizens and the city in the care and regeneration of urban commons" represents a key game changer in the city of Turin. By signing a so-called "Collaboration Pact" with the municipality, citizens are enabled to care for "common goods" such as public (and private) urban spaces, buildings, gardens, parks, and even schools. The agreement acknowledges active citizens as equal to the public administration by acting in the general interest. The regulation empowers local communities in managing co-created NBS, making them the actual keeper of the new urban spaces for the benefit of all citizens in the LL.

SPAZIOWOW represents a successful NBS intervention: Creation of a new public space located in the open area of an abandoned building that used to house the Institute for Agricultural Mechanization of the National Research Center. ProGIreg kick-started the undergoing transformation process and new vocation of reusing and opening it to citizens. The local community selected the location as a target area since it was indicated as "common good" to be revitalized for public use after years of disuse. The city of Turin managed the administrative procedures quickly given the external building structure's and surrounding open area's good condition.

142 Nature-Based Solutions for Urban Renewal in Post-Industrial Cities

The space allowed for experimenting with three NBS:

1. extensive green roof planted with natural grass;
2. pollinator-friendly garden in boxes with plants for pollinating insects; and
3. apiary for honey production.

The pollinator-friendly garden was designed to create an intimate, easily accessible space for all, equipped with garden boxes and benches and portals for multi-functional uses. During co-design, the local community decided the layout of the garden boxes and typology of flowers and vegetables to be planted. Moreover, Coldiretti (Federation of Italian Farmers) participated being a key stakeholder using the space for a local produce farm market. The co-design process also helped to form a "care group" of the space (formal and informal users of SPAZIOWOW taking care of the area and opening the area to all). Based on this, co-management of the garden in boxes was entrusted to a Collaboration Pact signed by the Community Foundation Mirafiori Onlus, Coldiretti, Parco del Nobile Association (association expert in urban farming) and an informal group of citizens "Rete Pollinatori Metropolitani", experts in beekeeping. The collaboration agreement stipulates open access to Mirafiori Sud residents while providing a quality green space to learn and experiment through Citizen Science laboratories and joint recreational activities.

The implementation at SPAZIOWOW faced COVID-19 restrictions and administrative barriers. The green roof meant to be publicly accessible for educational and recreational activities. However, the City of Turin Heritage Management Department excluded citizen access via the existing internal stairs due to transferring the use of the building to a new entity, prohibiting mixed use of the stairs. This misalignment in the revitalization of roof and building, exacerbated by rigid local regulation on heritage management, led to changing the concept to a non-accessible extensive green roof, thus limiting the NBS's full potential.

Case study: Dortmund – leveraging pollinator diversity with local communities

The NBS "Improving pollinator diversity" is an example of a co-created non-governmental-led NBS in the Dortmund LL. The local community-initiative co-developed the NBS interventions of creating pollinator-friendly meadows from the start. Given this NBS is largely non-technical, little technical expertise or complex planning procedures

Co-design NBS   **143**

besides botanical knowledge are required. These factors are particularly conducive for active citizen engagement due to few barriers.

The LL in the district Huckarde continues to suffer from the post-industrial legacy of underused open areas, both within the built environment and along the formerly heavily polluted and regulated river Emscher. Therefore, Huckarde citizens and the local NGO "Die Urbanisten" founded an association called "Naturfelder Dortmund e.V.", initiating sowing of wildflower meadows to increase pollinator-friendly vegetation while creating attractive green spaces for LL residents. The support of the well-embedded NGO, the "Naturfelder" initiative, acted as a catalyst in connecting different local stakeholders to implement the interventions. During the pandemic, the co-design process brought together the water management utility as the key landowner of the riverbanks along the river Emscher, wild flower meadows experts to advise on appropriate local seeds to attract endemic pollinator species. Despite lockdowns, the founding members of the initiative started transforming unused spaces in Huckarde by sowing wildflower meadows in spring 2021, accompanied by installing insect hotels to benefit insects.

Being embedded in the proGIreg project enabled the association to start its activities before its official founding. Direct communication between key stakeholders enabled a rather informal co-creation process not tied to lengthy bureaucratic decision-making and regulatory frameworks. A key difference of this case study is that the city of Dortmund has no direct involvement. However, it supports and promotes the project. The Dortmund Green Space Department follows its own strategy to increase pollinator biodiversity by identifying areas and creating wildflower meadows.

The initiative "Naturfelder e.V." has since generated high interest among diverse stakeholders, increasing numbers of enquiries by citizens to participate in the project. Private landowners have approached the association offering spaces for establishing wildflower meadows to promote pollinator diversity in areas that lack quality green spaces, within the LL and other parts of Dortmund. Thus, spiking replication and upscaling of the NBS as a grassroots movement and less a government-led strategy that empower citizens to provide quick solutions to local needs and improve living conditions and the quality of life.

Case study: Zagreb – co-managing a therapeutic garden

This case study presents a co-managed government model for co-creating NBS at the example of the LL activities in Zagreb. The municipality is a key stakeholder as principal landowner, while co-managing the urban community garden network throughout the city together with the local community who are engaged in maintaining and developing the offer in view of specific needs.

Urban gardening is embedded in Croatian culture and thus popular among citizens to grow food and socialize. A vivid urban garden community already exists in Zagreb's LL in the district of Sesvete. The growing population of Sesvete, which is characterized by many migrants following the Bosnian war in the 1990s, actively uses the existing urban gardens. In addition, the local NGO ZIPS plays an important role in raising awareness among citizens of protecting the environment and engaging residents in the urban regeneration of Sesvete using NBS including urban gardening and tree planting.

As part of transforming a post-industrial site in the Sesvete LL to strengthen its green infrastructure system, local citizens expressed the need for an extension of the existing urban garden to accommodate the district's high proportion of inhabitants with diverse special physical and mental needs. Dedicated local groups and initiatives mediated and facilitated communication between people with various disabilities, the municipality, and other relevant government bodies to respond to the requests for accessible gardening plots in the community gardens. Led by the city of Zagreb, the co-design process led to the idea of co-creating a specially designed therapeutic garden together with the local community to strengthen users' physical and mental health and contribute to social inclusion. Key features of the therapeutic garden are raised high growing beds that are accessible for wheelchairs and a sensory path.

The architecture faculty of the University of Zagreb developed possible scenarios for future development of the urban community garden together with students and local citizens. Given the specific requirements of the therapeutic garden to accommodate users with various special needs, a local landscape architecture bureau designed and planned the garden considering the input of therapists and other co-design outcomes. Construction of the therapy garden was taken over by the city-owned landscape firm. Shortly after completion, the garden was handed over to the local users who started planting the raised beds. The municipality, together with local groups working with disabled people are responsible for maintaining and managing the therapeutic garden in a co-management approach.

Unlike other forms of urban gardens that are often citizen-led bottom-up initiatives with little involvement of local authorities, the therapeutic garden in Zagreb Sesvete demonstrates how municipalities can successfully respond to specific local needs in partnership with the local community and users.

Co-design processes across different cities and EU H2020 sister projects

Co-creation of NBS for social inclusion in CLEVER Cities project in Milan

CLEVER Cities project has been developing a co-creation approach in three Front-Runner Cities (London, Hamburg, Milan) from June 2018 until September 2023. The co-creation process in each city differs given its local context and lead partners' responsibilities. Each city has set up Urban Innovation Partnership (UIP) and CLEVER Action Labs (CALs). This setup allowed ease of implementation within the co-creation process since the UIP worked as a centralized stakeholders group taking collective decisions.

The city of Milan developed NBS in three CALs: green roofs and walls across the city, a community park, and green noise barriers across a new train station (Mahmoud & Morello, 2020, 2023). The co-design process started in December 2019 across different social groups and used diverse co-design techniques and methods to ensure a wider citizen consensus in the three LLs:

- Co-design by immersion: workshop designed to collect stakeholders and citizens feedback within the physical location of the park itself that was co-designed later with the local community as in Giambellino 129 community park.
- Co-design using digital surveying: online digital survey on co-design of the CALs distributed in Tibaldi train stop with the support of a project facilitation team since the COVID-19 pandemic starting April 2020 till September 2020 has pushed the UIP to digitally engage with local communities.
- Co-mapping tool: implemented in the green roofs and walls CAL, which helped understand preferences and existence of such NBS measures in different parts in the city of Milan.

In addition, private partners and local authorities developed an awareness-raising campaign to push citizens' knowledge on green roofs and walls (see Mahmoud, Sejdullahu & Morello, 2021).

The Giambellino 129 community park co-design process adopted another approach for social inclusion: 2019, several workshops moved online during the pandemic in 2020. Observations and online surveys were conducted with two different groups: (1) limited number with key stakeholders on

146 Nature-Based Solutions for Urban Renewal in Post-Industrial Cities

inclusiveness in the executive design, and (2) group engaged on the larger district scale using online surveys to measure social engagement and cohesion in the project area at the pre-greening phase. Both methods proved to be comprehensively complete on providing evidence of inclusiveness within the co-design process and consequently decision-making mechanisms within the overall co-creation process (Mahmoud et al., 2021). Survey results showed key stakeholders involved included neighbourhood NGOs, the facilitation team, and a large group of residents.

Lastly, the third CAL encompasses a technical installation for a green noise barrier around the railway infrastructure in social housing neighbourhood Tibaldi-Giambellino in Milan's Southern transect. The co-design process was limited, as the public space adjacent to the station was expected to be used as a public waiting room for the train stop. Co-design online activities used visual surveying and distributing online questionnaires across two districts including the station. The results confirmed possible design choices, aesthetics, and types of NBS planned in the public space, hence be used for implementation based on a wider resident consensus.

At the end of the project, social monitoring focuses on questionnaires that measure post-greening impressions from implemented NBS and the possible increase of social cohesion and inclusion using common spaces created in the different CALs as per CLEVER Cities project objectives.

Co-creation of Edible City Solutions across different cities in EdiCitNet

The global movement of Edible Cities and of urban food initiatives uses a wide range of different governance forms, which are reflected in the approaches within the Edible Cities Network. Edible City Solutions (ECS) include all forms of urban agriculture and horticulture such as community and allotment gardens, but also forms of production, processing, and distribution of food in and on buildings (e.g., roof farming, aquaculture, indoor farming). Additionally, activities aim at transforming urban food systems towards more sustainable and resource-efficient circular economy and ecological design principles (Säumel et al., 2019) and urban food commoning such as urban food councils, food sharing, or food foraging (Scharff et al., 2019).

Given the movement's diversity, modes of governance are different (Table 8.1), which are also reflected in the approaches within the Edible Cities Network (Rotterdam (NL), Oslo (NO), Havana (CU), Andernach & Berlin

Co-design NBS   147

(DE), Montevideo (UY), Sant de Feliu de Llobregat (ES), Letchworth (UK), Carthage (TN), Sempeter pri Gorici (SI), Lomé (TG); Guangzhou (CN)).

In all cities, relevant stakeholders participate in an intentionally non-hierarchical organized city team as "community of knowledge and practice" to co-create LLs for showcasing the multifunctionality of ECS and co-design masterplans for anchoring ECS as strategic tools into urban planning (Esashika et al., 2023). There are cities with a clear top-down approach with government actor–led models. The public administration actively initiates and supports the integration of ECS and then encourage citizen participation. One top-down example is the Edible Andernach (Germany), where a municipal service company integrates and cares for edible elements such as fruit trees or vegetable beds into public green design. Based on this, the EU H2020 EdiCitNet project aims to encourage new actors to actively participate in projects with children and schools and thus transition from "nice to have" elements to elements anchored in the daily life of neighbourhoods. Havana (Cuba) developed a national programme to promote urban farming to secure the urban food supply since the economic crisis of the early 1990s. Valuable agricultural land is explicitly protected from urban development and leased to farmers, who are supported by a large network of knowledge, technologic innovation, and distribution channels. On the other hand, there are strong self-governing bottom-up movements, which originally partly opposed against administrative barriers into the urban tissue claiming their right to the city. However, city administrations have now recognized and value the multiple benefits of ECS for neighbourhoods, such as managing challenges, empowering neighbours, and promoting green economy and social cohesion in line with Berlin (Germany) and Rotterdam (Netherlands).

Conclusions

Co-design aims at embedding NBS into the broader sustainability management approach in regenerating post-industrial cities. To achieve this, collaboratively developing a productive green infrastructure system to empower local communities improves their living conditions. Active citizen and diverse stakeholder engagement early in co-creation processes ensures citizen ownership in the longer term. However, not all NBS are equally suited.

Three key and often interdependent factors are driving different degrees of co-creation engagement:

1) type of NBS considered for a specific purpose or post-industrial area, largely depending on the level of technical expertise and administrative procedures required;
2) underlying governance models strongly influence forms of citizen involvement whereby co-creation aims at shifting conventional government roles from leading to enabling; and
3) creating regulatory frameworks enabling citizens to take on responsibility for using public and private land in urban regeneration (e.g., Urban Commons Regulation of the City of Turin).

The co-creation processes in proGIreg and other EU H2020 projects highlight that complex NBS requiring lengthy bureaucratic procedures are successfully implemented by more government-led governance models. Local authorities act as key coordinator, supported by other local stakeholder groups. Intensity of stakeholder involvement may vary widely, notably in the planning and co-designing phase local community stakeholders are often less engaged but can take on greater responsibilities in co-managing and co-maintaining NBS under the guidance of the municipality and/or private-sector stakeholders given specific expertise. Many lower-tech NBS implemented in partner cities demonstrate that non-governmental-led governance models drive citizen engagement at all levels of the co-creation processes. Replicating NBS in proGIreg's Eastern European Follower cities also highlighted the fact that co-creation processes are new forms of collaboration between local authorities and civil society stakeholders.

Transparent co-design processes identify and flag technological and non-technological barriers of NBS implementation in each context to maximize the impact of upscaling and replicating NBS under any local conditions. This is not to say that embedded systemic planning cultures and supporting policies are needed more than ever to foster lively stakeholder engagement, not least in ensuring co-maintenance of NBS for long-term sustainability.

Notes

1. Including policies, guidelines, or procedures that are not favourable for implementation and upscaling; insufficient legislation and policies that would facilitate procedures, challenges linked to government assistance or political support, unfavourable planning schemes and more.

2. Including human or society induced challenges and constraints that are originating from social norms and/or cultural values; they may also refer to education, awareness, capacity building, stakeholder management and priorities, social inclusion and cohesion issues and more.
3. Including constraints to entry in financial market, lack of funding, lack of mainstreaming processes for NBS that will bring the necessary funding, inadequate or ineffective financing schemes, unsustainable funding processes.

References

Bradley, S., Mahmoud, I. H., & Arlati, A. (2022). Integrated Collaborative Governance Approaches towards Urban Transformation: Experiences from the CLEVER Cities Project. Sustainability 2022, 14(23), 15566. https://doi.org/10.3390/SU142315566

Esashika, D., Masiero, G., & Mauger, Y. (2023). Living labs contributions to smart cities from a quadruple-helix perspective. Journal of Science Communication, 22(03). https://doi.org/10.22323/2.22030202

Frantzeskaki, N., Mahmoud, I. H., & Morello, E. (2022). Nature-Based Solutions for Resilient and Thriving Cities: Opportunities and Challenges for Planning Future Cities. In Mahmoud, I. H., Morello, E., Lemes de Oliveira, F., & Geneletti, D. (Eds.), *Nature-based Solutions for Sustainable Urban Planning. Contemporary Urban Design Thinking*. Cham: Springer. https://doi.org/10.1007/978-3-030-89525-9_1

Hölscher, K., Frantzeskaki, N., Lodder, M., Allaert, K., & Notermans, I. (2020). Coproduction Guidebook, Connecting Nature, H2020 Grant no. 730222. Retrieved from https://connectingnature.eu/sites/default/files/downloads/First%20version%20Co%20Production%20guidebook%2030%20Aug%202020.pdf

Latinos, V. (2020). Report on Non-technological Barriers, Del. 5.3, proGIreg. Horizon 2020 Grant Agreement No 776528, European Commission. Retrieved from https://progireg.eu/resources/NBS-business-models/

Mahmoud, I., & Morello, E. (2020). Are Nature-based Solutions the Answer to Urban Sustainability Dilemma? The Case of CLEVER Cities CALs within the Milanese Urban Context. *Atti della* XXII Conf. Naz. SIU. L'Urbanistica Ital. di Front. all'Agenda 2030. Portare Territori e comunità sulla Strada della sostenibilità e della resilienza, pp. 1322–1327.

Mahmoud, I., & Morello, E. (2021). Co-creation Pathway for Urban Nature-based Solutions: Testing a Shared Governance Approach in Three Cities and Nine

Action Labs. In Bisello, A., et al. (Eds.), Smart and Sustainable Planning for Cities and Regions, Green Energy and Technology (pp. 259–276). https://doi.org/10.1007/978-3-030-57764-3_17

Mahmoud, I., & Morello, E. (2023). Four Years of Co-creation with Stakeholders: What Did We Learn about Its Added Value in Urban Planning? Insights from CLEVER Cities Milan Three Urban Living Labs. In Cerreta, M. & Russo, M. (Eds.), La valutazione come parte del processo pianificatorio e progettuale, Atti della XXIV Conferenza Nazionale SIU Dare valore ai valori in urbanistica (pp. 76–85). Planum Publisher e Società Italiana degli Urbanisti. https://media.planum.bedita.net/00/7a/Atti%20XXIV%20Conferenza%20Nazionale%20SIU_Brescia_VOL.09_Planum%20Publisher_2023_.pdf

Mahmoud, I. H., Morello, E., Vona, C., Benciolini, M., Sejdullahu, I., Trentin, M., & Pascual, K. H. (2021). Setting the Social Monitoring Framework for Nature-Based Solutions Impact: Methodological Approach and Pre-Greening Measurements in the Case Study from CLEVER Cities Milan. Sustainability, 13(17), 28. https://doi.org/10.3390/su13179672

Mahmoud, I. H., Sejdullahu, I., & Morello, E. (2021). Milan's ULL Co-design Pathway to Spread Green Roofs and Walls throughout the City. In Digital Living Lab Days 2021 (Ed.), *Change the Future Together: Co-creating Impact for More Inclusive, Sustainable & Healthier Cities and Communities* (pp. 288–295). Retrieved from https://openlivinglabdays.com

Säumel, I., Reddy, S. E., & Wachtel, T. (2019). Edible City Solutions—One Step Further to Foster Social Resilience through Enhanced Socio-Cultural Ecosystem Services in Cities. *Sustainability*, 11(4), 972. https://doi.org/10.3390/su11040972

Scharf, N., Wachtel, T., Reddy, S. E., & Säumel, I. (2019). Urban Commons for the Edible City—First Insights for Future Sustainable Urban Food Systems from Berlin, Germany. *Sustainability*, 11, 966. https://doi.org/10.3390/su11040966

Tandon, R., Singh, W., Clover, D., & Hall, B. (2016). Knowledge Democracy and Excellence in Engagement. *IDS Bulletin*, 47(6). https://doi.org/10.19088/1968-2016.197

Voorberg, W. H., Bekkers, V. J. J. M., & Tummers, L. G. (2015). A Systematic Review of Co-Creation and Co-Production: Embarking on the Social Innovation Journey. *Public Management Review*, 17(9), 1333–1357. https://doi.org/10.1080/14719037.2014.930505

Voytenko, Y., McCormick, K., Evans, J., & Schliwa, G. (2016). Urban Living Labs for Sustainability and Low Carbon Cities in Europe: Towards a Research Agenda. *Journal of Cleaner Production*, 123, 45–54. https://doi.org/10.1016/j.jclepro.2015.08.053

Wilk, B. (2020). Guidelines for co-designing and co-implementing green infrastructure in urban regeneration processes, Del. 2.10, proGIreg. Horizon 2020

Grant Agreement No 776528, European Commission, retrieved from https://progireg.eu/fileadmin/user_upload/Deliverables/D2.10_Guidelines_for_co-designing_proGIreg_ICLEI_200804.pdf

Wilk, B., Säumel, I., & Rizzi, D. (2021). Collaborative Governance Arrangements for Cocreation of NBS. In Croci, E., & Lucchitta, B. (Eds.), Nature-based Solutions for More Sustainable Cities: *A Framework Approach for Planning and Evaluation*, Emerald Publishing Limited, Leeds, pp. 125–149. https://doi.org/10.1108/978-1-80043-636-720211012

Zingraff-Hamed, A., Huesker, F., Lupp, G., Begg, C., Huang, J., Oen, A., ... & Pauleit, S. (2020). Stakeholder Mapping to Co-create Nature-based Solutions: Who Is on Board? *Sustainability*, 12(20), 8625.

Turin Living Lab   9

Silvia Barbero and Federica Larcher

The Turin Living Lab

Living Labs (LLs) have gained significant popularity in urban settings, emerging as an innovative mechanism for testing, validating, and refining nature-based solutions (NBS) in real-life environments. This chapter delves into the comprehensive exploration of the LL situated in the Mirafiori Sud district of Turin, Italy, one of the three European Front-Runner Cities (FRCs).

Turin is a significant business and cultural hub in Northern Italy, being the capital city of the Piedmont region and the Metropolitan City of Turin. Situated primarily on the western bank of the Po River, below the Susa Valley, the city is embraced by the western Alpine arch and Superga hill. Turin has a population of 843,514 (2023), the broader urban area is home to around 1.7 million inhabitants, estimated by Eurostat.

Turin has undergone a transformative journey since the 1990s, shifting from an automotive industrial center to a thriving city fostering start-ups and business innovation. Notably, this transformation catapulted Turin to the runner-up position in the 2016 'European Capital of Innovation' competition and to be 'Design Capital' in 2018. Turin's commitment to sustainability is underscored by its recent initiatives, including the establishment of extensive networks of parks, green cycling lanes, and verdant corridors along rivers and former railway lines. Consequently, Turin boasts more green space per inhabitant than any other Italian city, contributing to a holistic approach to urban development.

The proGIreg LL is located in the Mirafiori Sud district, nestled along the banks of the river Sangone. Historically a working-class area with a

DOI: 10.4324/9781003474869-9

This chapter has been made available under a CC-BY license.

Turin Living Lab **153**

population of more than 30,000 and diverse social groups, Mirafiori Sud holds immense potential for urban regeneration. The district is characterized by active local associations, a robust cultural heritage, and vacant industrial buildings that serve as canvases for new community initiatives. Mirafiori Sud, with its dynamic blend of community engagement, cultural richness, and adaptive urban spaces, stands as a compelling example of how LL can be instrumental in shaping sustainable and vibrant urban futures.

The Living Lab in post-industrial Mirafiori Sud district

Mirafiori Sud encompasses a substantial portion of the city's southwestern suburbs. Historically known as the city's Circoscrizione 10, it underwent administrative changes in 2016, merging with Circoscrizione 2 as per the Municipal Decentralization Regulation No. 374/2016.

Other key aspects of Mirafiori Sud LL in Turin are listed in Table 9.1.

The Mirafiori Sud district stands at a crossroads, presenting both vulnerabilities and extraordinary opportunities that hold the potential to redefine the area's landscape through the successful implementation of various NBSs. The LL encapsulates distinctive features that shape the district's narrative: the district's essence is woven with robust community ties, yet the emergence of disconcerting dynamics (i.e., economic decline, urban decay, social issues, demographic changes, housing issues) poses a threat to this social cohesion. In its favor, the presence of community foundations and citizen associations acts as a buffer, preventing further degradation of the social fabric at the local level (Barbero et al., 2022).

In the face of the multiple challenges and opportunities of Mirafiori Sud, a particular problem emerges due to the interaction between the low

Table 9.1 Key characteristics of the Turin Living Lab in Mirafiori Sud

Key characteristics/challenges	Description
Size	1149 ha
Population	c. 34,000
Environment	Transforming brownfields and underused spaces into community gardens
Socio-economic profile	Residents with social problems, high population density, and with a neighborhood-dormitory isolated from surrounding areas and with poor services

population density and the vast expanse of empty space caused by the region's industrial decline. This hinders meaningful interactions, impedes communication, and disrupts connections between citizens, companies, and associations.

Key challenges of the LL include surging health issues, notably cardio-respiratory ailments and chronic mental stress, partly due to demographics characterized by a high proportion of single elderly individuals facing psychological discomfort. In addition, the LL suffers from disquieting youth unemployment rate, surpassing the alarming threshold of 50%, coupled with a general trend of lower education levels. The economic landscape mirrors this imbalance, evident in the scarcity of local businesses, as the predominant workforce finds employment within the service sector.

NBS interventions in proGIreg sought to address this complex social fabric.

Amidst these challenges lies a dual opportunity. The district is marked by low real estate values and a surplus of vacant accommodations, embodying both a hurdle and an alluring prospect. While these factors pose a formidable challenge to be addressed, they also serve as a force, potentially enticing new citizens to the district's transformative potential. The city of Turin as a key stakeholder in experimenting with NBS in the proGIreg project is committed to sustainability and envisions a harmonious coexistence between urban development and the natural environment.

Figure 9.1 Mirafiori.

Source: City of Turin

Policy frameworks impacting the Turin Living Lab

The widespread adoption of NBS seamlessly aligns with overarching frameworks such as the blue-green infrastructure paradigm. NBS emerge as a multifaceted strategy with profound implications for the economy, environment, and society. In urban and post-industrial landscapes where sustainability integration is paramount, NBS stands as a catalyst for positive change. Beyond providing measurable economic benefits to both citizens and entrepreneurs, it plays a pivotal role in enhancing climate resilience, aligning with the Green Deal adaptation goals and the Global Biodiversity Framework (GBF) target areas. The Operational Programme 2014–2020 (OP) for the Piedmont Region is structured around key pillars, featuring the Green Economy and Sustainability theme and connected to three priority axes (PAs):

- **PA1 (Research and innovation [R&I]):** Envisioning bolstering R&I infrastructure and fostering stronger ties between research and industry, with a primary focus on developing innovative products and services, incorporating social and eco-innovation.
- **PA3 (Competitiveness of SMEs):** Targeting the enhancement of competitiveness in sectors like agriculture, fishing, and aquaculture, emphasizing new business models and green economy principles. Concurrent investments in greening SME processes are slated.
- **PA4 (Low-carbon economy):** Advocating for the promotion of renewable energies and energy efficiency within companies, identifying circular economy processes as potent contributors.

Turin secured a landmark position among the 100 European Mission cities committed to achieving zero greenhouse gas emissions by 2030, emerging as a hub for innovation in the climate services field. In this context, the Turin LL continues to lead experimentation and testing of services linked to ecological transitions, affirming its role as a driving force behind the city's progressive and sustainable evolution.

In total, 17 NBS have been implemented in the LL, creating extensive collaborative networks of local stakeholders and local communities extending beyond official proGIreg partners. The initial plenary meeting with diverse stakeholders underscored the need for a multifaceted approach to manage planning activities effectively. Recognizing this, three distinct 'boards' were established to oversee specific aspects of the proGIreg planning process. A proGIreg partner was appointed as coordinator for each board, and a 'core team' was formed to foster cohesive planning among local partners. The City of Turin, under the European Funds and Innovation Department, assumed the role of overall coordinator, employing a 'variable geometry' principle to ensure dynamic and tailored engagement in proGIreg's planning endeavors.

NBS in Turin Living Lab

In Turin, seven of the eight NBS types have been applied and monitored aimed at the redevelopment of the former industrial neighborhood Mirafiori Sud to promote a multifunctional urban green infrastructure aimed at improving the quality of citizens' lives, to reduce climate vulnerability, and to increase biodiversity in the LL. Crucially, many schools in Mirafiori Sud actively participate in implementing and testing innovative solutions proposed by the LL. This inclusive approach empowers local families and citizens to become integral contributors to the development and transformation of their own neighborhood.

Figure 9.2 shows each NBS location within the Turin LL. Each NBS realized in the city has been co-designed in continuous discussions with local stakeholders including public, private, academic, and non-profit stakeholders according to the quad helix approach of proGIreg. This participatory approach has proven a key tool for sustainable co-creation in response to environmental quality and social inclusion objectives.

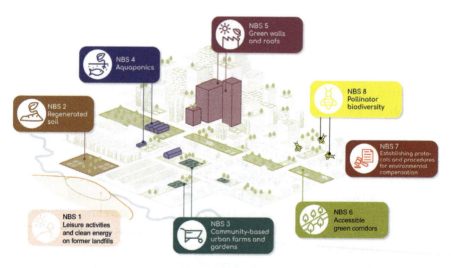

Figure 9.2 Seven of proGIreg NBS implemented in the Turin Living Lab.
Source: RWTH

In the following, the implemented NBS are briefly described below, demonstrating the roles of different stakeholders involved and the impact on the LL.

NBS 2 – New soil (new forest on regenerated soil): The initial idea stems from the city of Turin's need for local fertile soil for the creation of new parks and green areas. On the other hand, huge amounts of excavated earth and rocks, removed from construction sites, are always available, but without more land consuming. In the park along the Sangone river, a regenerated technosoil was used, obtained by mixing rocks materials, compost, zeolite, and mycorrhizae, a biotic compound to stimulate the growth of plants. The Municipality and Envipark, Environment and Technology Research Center developed the concept for this NBS c the DUAL s.r.l., a company that operates in the excavated earth sector; and Acea Pinerolese, a company that produces compost. This resulted in experimenting with a regenerated technosoil at a site in the park along the Sangone river. The regeneration process entailed mixing rocks materials, compost, zeolite, and mycorrhizae, a biotic compound to stimulate plant growth. The University of Turin, Department of Agricultural, Forest and Food Sciences and Department of Chemistry, as scientific consultants, guided the design choices for the regenerated soil and the plants. The experimental urban forest has been designed with the aim of evaluating the capacity of ornamental plant species to grow in the regenerated soil, using species commonly used in urban greening while testing new ones, considering low maintenance needs and high resilience to extreme climate conditions. The area (2,000 m^2) was

Figure 9.3 Turin Living Lab map locating and describing NBS interventions.
Source: ProGIreg

divided in two plots, one with the new soil and the other with the original soil. Five tree species *Celtis australis, Gleditsia triacanthos* 'sunburst', *Malus* × *evereste, Quercus ilex,* and *Tilia cordata* 'greenspire', and five shrubs species *Eleagnus ebbingeii, Physocarpus* 'diable d'or', *Spirea vanhouttei, Teucrium fruticans,* and *Ligustrum texanum.* The University of Turin was involved in monitoring the site for five years, collecting chemical, environmental, and biological data (Ascione et al., 2021) to assess the viability of the innovative NBS. This NBS offers a robust business model; however, for replication purposes it requires a specific stakeholder setup that may not be obvious or available in every city.

NBS 3 – Urban community gardens: Many vegetable gardens have been created throughout the neighborhood, both soil-bound and soil-less in boxes, involving citizens, scholars, and notably vulnerable groups, led by two active NGOs in the district: Orti Generali and Fondazione Mirafiori.

The most successful NBS intervention is managed by the Orti Generali association, in charge of a community urban garden on public land (2,5 ha) owned by the city of Turin (Figure 9.3). Orti Generali has achieved social inclusion and environmental sustainability on formerly unused land. In total, 160 vegetable gardens were created and more than 300 autochthonous trees planted. The innovative concept is based on a new model of urban gardening with rents for plots depending on income or voluntary shared responsibilities. The gardens are assigned to private citizens, families, and poor people. Being a social enterprise, the garden also offers training and job placement initiatives to improve the safety and social aggregation. The educational area with an urban farm, beehives, a greenhouse classroom, and some community gardens complete the NBS while being important for biodiversity conservation (NBS 8) thanks to pollinator-friendly areas. Since 2018 the monitoring of Lepidoptera, Apoidea (wild bees), and of the flora useful to pollinating insects was done, and results were shared with the gardens users.

NBS 4 – Aquaponics: The city embarked on a visionary initiative by championing aquaponics experiments, not merely as isolated endeavors but as dynamic showcases designed for widespread dissemination. This multifaceted undertaking sought to serve several purposes, aligning with the city's commitment to innovation and sustainable practices. First and foremost, the initiative aimed to test and evaluate cutting-edge technologies integral to the innovative fusion of fish farming and vegetable production known as aquaponics. By serving as a testing ground for these emerging technologies, the city fostered an environment where experimentation and refinement could take place, paving the way for advancements in sustainable agriculture. The aquaponics experiments go beyond technological testing

Turin Living Lab 159

Figure 9.4 Urban gardens in Mirafiori Sud.

Source: Politecnico di Torino

by evolving into a dynamic platform for learning, community involvement, and the progressive integration of sustainable practices into the urban landscape.

NBS 5 – Green roofs and walls (Radić, 2019): The NGO Orti Alti created an extensive green roof on a public building. In the surrounding court was also created a garden for pollinators (Orto WOW), and beehives were installed. Furthermore, two green walls (one outdoor and one indoor, Figure 9.4) have been implemented in a homeless shelter and at a primary school. The implementation was coordinated by the city executed by Verdeprofilo company and scientific support from the Politecnico di Torino. Some environmental monitoring was conducted by a proGIreg third-party partner, while the University of Bari evaluated the perception of nature of children before and after the green wall installation at the school.

NBS 6 – Green corridors: A transformative green corridor has been co-designed and brought to life, serving as a conduit to enhance connectivity between the park along the Sangone river and the rest of Mirafiori Sud. This not only fosters a seamless link between these two distinct locales but also acts as a catalyst for the harmonious integration of nature into

Figure 9.5 Green wall at the shelter for homeless people.

Source: Politecnico di Torino

the heart of Mirafiori Sud, facilitating the flourishing of vital pollinator communities within the urban landscape. This green corridor represents a passageway and a deliberate effort to invite nature into the city, creating a vibrant and ecologically enriched environment. Beyond its function of connecting spaces, this corridor is a testament to the city's commitment to biodiversity and sustainable urban development. A meandering walking path now weaves through clusters of carefully curated green spaces offering residents and visitors encounters with nature. Such urban planning with NBS prioritizes ecological harmony, community well-being, and the creation of inviting spaces for recreation. Mirafiori Sud's newest green artery has significantly enhanced the urban quality.

NBS 7 – Strategic public-private partnership for greening the city: This intangible NBS aims at designs catalysing the transformation of urban spaces through the promotion of local initiatives to greening the city. The city has taken a proactive approach by presenting a comprehensive catalogue of environmental actions, strategically tailored to align with corporate social responsibility initiatives. This catalogue serves as a valuable resource, offering companies a spectrum of environmentally conscious actions that resonate with their commitment to fostering a sustainable and eco-friendly footprint. Among the diverse range of suggested actions, noteworthy is the urban forestation along the scenic Sangone river. This initiative is conceived to be financed by private companies, marking a collaborative synergy between public and private sectors in the pursuit of a shared environmental vision and the enhancement of the city's green infrastructure. By investing in the growth of green spaces along the Sangone river, these companies bolster the city's ecological resilience and create aesthetically pleasing environments that contribute to the overall well-being of the community. Thus, acting as a catalyst for positive change within the urban landscape and exemplifying a forward-thinking approach that empowers local initiatives and forges collaborative partnerships to create a sustainable, green, and vibrant urban environment for both present and future generations.

NBS 8 – Pollinators biodiversity actions: The scientific approach implied the citizen involvement in creating and monitoring the spaces for the benefit of the pollinators. In Mirafiori Sud, the project 'Butterflies in ToUr' created by the Department of Biology of the University of Turin used an approach that is designed to be socially inclusive and scientific, thanks to the collaboration of researchers and citizens with doctors and patients of mental health centers for the promotion and management of pollinator-friendly green areas.

Challenges and opportunities

Significant challenges and opportunities of co-creating NBS are shaping the trajectory of the proGIreg project. Noteworthy achievements and ongoing challenges underscore the dynamic nature of working with NBS in funded research project:

1. **Citizen co-ownership for vibrant installations**
   - **Achievement:** The proactive involvement of citizens has been pivotal in breathing life into NBS installations, fostering a sense of co-ownership that mitigates the risk of vandalism.
   - **Challenge:** Sustaining citizen engagement remains a continual challenge, demanding innovative strategies to ensure ongoing co-ownership and active participation.
2. **Orti Generali community garden as a catalyst**
   - **Achievement:** The successful implementation of the community garden Orti Generali has not only served as a hub for urban agriculture within the LL but has also drawn gardening enthusiasts from other Turin districts to this often-overlooked area.

162 Nature-Based Solutions for Urban Renewal in Post-Industrial Cities

- **Challenge:** Maintaining the momentum and expanding the reach of such community-driven initiatives presents an ongoing challenge.

3. **Adapting to COVID-19 restrictions**
   - **Achievement:** Despite severe disruptions caused by COVID-19 restrictions, the various NBS teams showcased resilience by shifting to online platforms for co-design, engagement, and maintenance activities.
   - **Challenge:** Navigating the evolving landscape of virtual engagement requires ongoing creativity to ensure effective stakeholder communication and project progression.

4. **Navigating bureaucratic obstacles for co-implementation**
   - **Achievement:** The co-implementation process encountered bureaucratic hurdles, providing valuable insights into navigating public procedures. ProGIreg has facilitated a nuanced understanding of these challenges, leading to innovative regulatory approaches.
   - **Challenge:** Ongoing complexities in bureaucratic processes necessitate continuous efforts to streamline procedures and foster collaborative agreements between the public and private sectors.

The interplay of achievements and challenges reflects a commitment to dynamic, community-driven urban transformations. As the project evolves, each hurdle becomes an opportunity for innovation, resilience, and collaborative problem-solving, contributing to the sustainable development of Turin's urban landscape.

Impact of the COVID-19 pandemic in Turin Living Lab

The onset of COVID-19 and the subsequent imposition of restrictions impaired proGIreg project activities significantly, resulting in severe delays in co-design, implementation, and maintenance activities. The pandemic-induced limitations compelled a re-evaluation of traditional approaches to stakeholder engagement and beneficiary interaction. In the face of physical distancing measures the diverse teams overseeing various NBS displayed commendable resilience and adaptability. Innovative measures utilizing digital platforms and tools to bridge the physical gap sustained meaningful contact with stakeholders and beneficiaries.

These virtual endeavors also emerged as platforms for continued collaboration and knowledge exchange. Through webinars, virtual workshops, and online consultations, the NBS teams fostered ongoing dialogue, ensuring that the perspectives, insights, and needs of stakeholders and beneficiaries remained integral to the project's trajectory. The ability to seamlessly transition to online engagement underscored the commitment of the proGIreg project to maintaining its participatory ethos despite external challenges. By leveraging technology and creativity, the project not only weathered the storm of disruptions but also emerged with strengthened connections and a deeper appreciation for the role of digital platforms in facilitating meaningful collaboration in times of adversity.

Lessons learnt

Over the five years of proGIreg project in the Turin LL, the co-design experiences, employing the quadruple helix approach have been instrumental in fostering public participation and awareness (Baccarne, 2014). The experiences of experimentation with NBS such as green walls have engaged citizens, hence emphasizing the valuable role of regeneration processes in natural ecosystems for enhancing social cohesion. A critical point remains concerning the resilience of co-design results over the long term, the reliability and scalability of experiments in different LLs, and their economic sustainability. These aspects remain subjects for ongoing study and raise the question of securing continued maintenance at the early co-design stages without compromising flexibility.

The efficacy of the adopted LL approach has been substantiated through tangible upscaling and replication outcomes. The NBS design methodology, honed through this experience, has extended its impact to other noteworthy European programme Urban Innovative Actions projects such as ToNite and the Fusilli focusing on urban food planning. This expansion of the approach has validated its versatility and amplified awareness of using of NBS in urban contexts.

ProGIreg's successes now extend to broader municipal initiatives. Drawing on the insights gained, the city of Turin has embraced the Strategic Plan for Green Infrastructure in 2020, a strategic blueprint actively promoting NBS and ecosystem services approach, positioning Turin as a resilient city that prioritizes sustainable urban development.

The aspiration is that NBS will transcend their current status to be recognized as pivotal to seamlessly integrating ecology and planning. Beyond creating a pleasant and healthful environment for inhabitants, NBS are poised to elevate biodiversity levels and implement tangible adaptation actions. This holistic perspective underscores the transformative potential of NBS, envisioning a future where these initiatives are indispensable components of urban development strategies. The fallout was to raise further awareness about the need of using NBS in urban contexts, also as monitoring stations for environmental quality (Larcher et al., 2021).

Crucially, the success of NBS implementation hinges on robust engagement processes. The active involvement and heightened awareness among residents are essential ingredients, fostering a sense of identification with their surroundings. This, in turn, stimulates a collective responsibility for the city's well-being and nurtures a profound spirit of community. NBS has the potential to shape urban landscapes while cultivating a harmonious coexistence between human activities and the natural environment, creating resilient, vibrant, and community-centric urban spaces.

References

Ascione G. S., Cuomo F., Mariotti N., & Corazza L. (2021). Urban Living Labs, Circular Economy and Nature-Based Solutions – Ideation and Testing of a New Soil in the City of Turin Using a Multi-stakeholder Perspective. *Circular Economy and Sustainability*, 1(2), 545–562. https://doi.org/10.1007/s43615-021-00011-6

Baccarne B., Mechant P., Schuurman D., Colpaert P., & De Marez L. (2014). Urban Socio-technical Innovations with and by Citizens. *Interdisciplinary Studies Journal*, 3(4), 143–156.

Barbero S., Giraldo Norha C., & Campagnaro C. (2022). Systemic Solutions for the Holistic Well-being of Cities. Processes, Results and Reflections. *Agathon*, 11, 50–61. https://doi.org/10.19229/2464-9309/1142022

Larcher F., Baldacchini C., & Calfapietra C. (2021). Nature-Based Solutions as Tools for Monitoring the Abiotic and Biotic Factors in Urban Ecosystems. In: Catalano C., et al. (eds.), *Urban Services to Ecosystems. Future City*, vol. 17, pp. 131–150. Cham: Springer.

Radić M., Dodig M. B., & Auer T. (2019). Green Facades and Living Walls – A Review Establishing the Classification of Construction Types and Mapping the Benefits. *Sustainability*, 11(17), 1–23. https://doi.org/10.3390/su11174579

Ningbo Living Lab 10

Ruowen Wu, Tian Ruan and Yaoyang Xu

The Ningbo Living Lab as testing ground for NBS

Ningbo is a sub-provincial division in northeast Zhejiang Province in the People's Republic of China, which is of similar status to a prefecture-level city (Figure 10.1). Before the COVID-19 pandemic, in 2019, the city had a registered population of 6.085 million, and the regional gross domestic product (GDP) was 118.5 billion yuan, an increase of 6.8% over the previous year. Ningbo has an elevation of about 5 meters and belongs to the north subtropical monsoon climate. As the city could be considered as a typical case of rapid urbanization in the eastern coastal region of China, Ningbo is facing many challenges concerning green and blue areas. Specifically, the soil and surface water of Ningbo were contaminated in quality and reduced in quantity due to the construction and spread of gray infrastructure (i.e., for transportation and buildings). The major contamination sources to Ningbo surface water included road pollution and waste water with heavy metals from heavy industries (Xu et al., 2023). Industrial and agricultural production including excessive input of fertilizer and pesticide, waste water of polluting industry, waste gas of mineral exploitation and coal combustion were important sources of soil contamination (Xiang et al., 2020). Given the challenges concerning green and blue areas, Ningbo has been listed as one of the pilot cities participating in a series of action plans launched by national and provincial governments, in order to address a series of soil and water challenges.

DOI: 10.4324/9781003474869-10

This chapter has been made available under a CC-BY license.

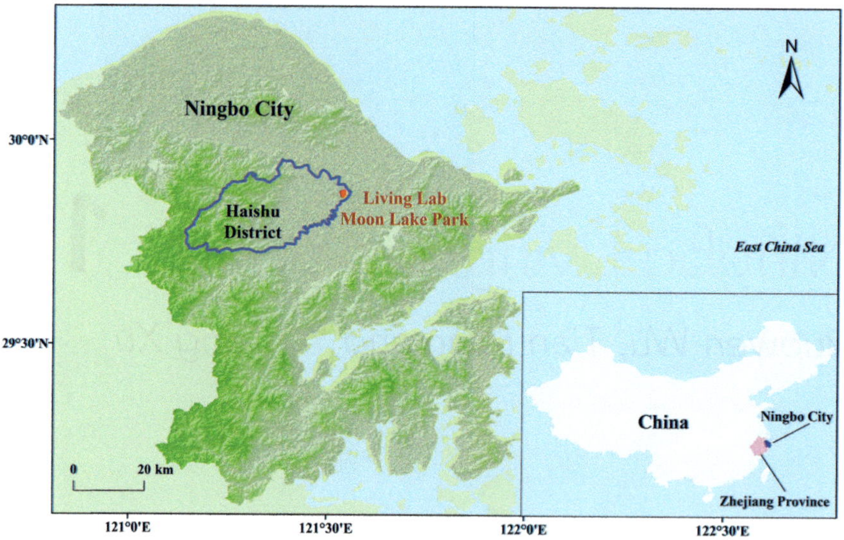

Figure 10.1 Location of the Ningbo city in China.

The Living Lab area in Haishu district, Ningbo

The Living Lab (LL) (2.07 km²) of Ningbo is the entire Moon Lake Street where the Moon Lake is located (Figure 10.2). Moon Lake Street is located in the downtown area of Ningbo city, with an area of only 2.07 km². It has jurisdiction over seven communities, with a population of 25,750 people and a density of 12,440 inhabitants (inh.)/km². In 2017, the green area of Ningbo city was 11.89 m²/inh., and the green area of Moon Lake Street was about 11.5 m²/inh., which was lower than China's per capita park green area of 14.01 m²/inh. Therefore, Ningbo's green infrastructure construction needs to be continuously strengthened.

In terms of social and cultural inclusion, although the population of Moon Lake Street and its seven communities has witnessed a decreasing trend, it is still very dense. Considering that the Street also accommodates the 28-hectare park located in its center, its density of over 12,400 inh./km² (2017) represents an important defining characteristic. Providing environmental services and nature-based solutions (NBS) to ensure quality of life could be a challenge in such a situation. Currently, there are many primary and secondary schools, theaters, large leisure sports venues, and museums, with the area being an attractor at district level. The area is very accessible via subway, and it is convenient to meet people's needs for education and culture. In terms of human health and well-being, Moon Lake Park located

Ningbo Living Lab 167

Figure 10.2 Location of the LL area in the Haishu district of Ningbo city.

in the center of Moon Lake Street is a municipal conservation zone for history and culture in Ningbo with a large number of attractions and leisure facilities around, which has contributed significantly to well-being. Besides, the residents of seven communities can easily access the green space for walking and entertainment, which has contribution to both physical and mental health.

In terms of ecological and environmental restoration, there is a reconstruction project called "Shi qing hu xi" on Moon Lake Street located on the west side of the lake, north to Zhongshan Road, west along Changchun Road, east to Haoyue Street, Gongqing Road, and south to Guijing Street. The construction scale of the "Shi qing hu xi" project is about 0.2 km², and the buildable area represents around 0.163 km². The Chinese-protected area is about 0.027 km², and the historic building is 0.039 km². However, Moon Lake Park is located in the urban area, surrounded by many old neighborhoods, with high land prices and large transformation costs. In terms of the economy and the labor market, Moon Lake Street is a mature tourist area with many hotels and restaurants, providing a large number of jobs for the labor force. Currently, Moon Lake Street is dominated by the service industry, and its industrial structure is relatively simple, which is vulnerable to market shocks. But nowadays, Chinese people are paying more attention to the quality of life as seen in the growing popularity of tourism and leisure, which provides support for the economic growth of Moon Lake Street.

168 Nature-Based Solutions for Urban Renewal in Post-Industrial Cities

Figure 10.3 Living Lab area in Ningbo.

Policy context impacting the Ningbo Living Lab

Due to rapid economic development, population growth, industrialization, and urbanization, the overall soil quality of Ningbo has shown a downward trend and thus facing challenges. In particular, the risk of heavy metal pollution still exists since heavy industry still occupies a major position in Ningbo's industrial structure. At the same time, organic pollution is prominent, and the pollutants are gradually migrating to agricultural products and water bodies. Problems such as insufficient development and utilization of contaminated sites have caused serious threats to agricultural products and human health and will affect social harmony and stability. In 2013, the Ningbo government issued a clean soil action plan to strengthen the comprehensive remediation of soil pollution sources, soil pollution monitoring and control, and pollution site remediation. When considering that these action plans are dominated by diverse governmental agencies, however, the key question concerning local government and multiple stakeholders is how top-down plans for regeneration of the existing green and blue infrastructure associated with intensive investments can be implemented effectively at local level. There is an urgent need for the transfer of transdisciplinary research to the top-down co-actions and the city-level co-practices when implementing NBS in partnership with local communities in Ningbo.

Although Zhejiang's water resources per unit area can rank fourth in China, 80% of water resources are distributed in mountainous areas, so eastern Zhejiang (including Ningbo), where the population is concentrated and the economy is developed, is a key area of water scarcity. In addition, there are four outstanding challenges in Zhejiang water resources, including the large gap between supply and demand, prominent structural contradictions, serious pollution, and low effective utilization. Therefore, in 2013, the Zhejiang Provincial Party Committee proposed the introduction of the "Five Water Treatment" to transform and upgrade water management by controlling sewage, preventing flood, draining flooded fields, guaranteeing water supply, and emphasizing water conservation. Ningbo has been at the forefront of urban river management in China. Also facing the four challenges, Ningbo has taken the lead in adopting the PPP (public-private partnership) management model to comprehensively harness and conserve the city's rivers. The black water bodies in urban rivers have been completely eliminated, and the river management in Ningbo has reached the stage of ecological restoration.

Challenges and goals of implementing NBS in Ningbo Living Lab

There are many problems associated with Moon Lake Street such as aging infrastructure, insufficient modern facilities, and difficulties in coordinating interests across social groups. Although Moon Lake Street is prosperous and full of tall buildings, there are many old residential quarters, old buildings, old streets, and old markets behind the high-rise buildings, with aging equipment and facilities and many remaining problems. Ten out of the 11 residential quarters in the streets are old ones that have been built for more than 15 years. In addition, there have been perennial outbreaks of algae in the Moon Lake in recent years, and some polluted water bodies with seasonal stench have appeared, which has affected the life and leisure quality of residents, and also seriously affected the image of Ningbo city and the beauty of Tianyi pavilion, the Moon Lake Scenic Area. Therefore, it is urgent to improve the water quality of the Moon Lake.

The goals of LL are in three aspects:

1) After the completion of the lake ecological comprehensive control project, the main water quality indicators should reach IV class within one year and reach III class 1 for two years.
2) Water quality purification and ecological restoration projects should continue to remove pollutants in water bodies through moderate human intervention; self-purification ability of water bodies should be improved by applying ecological technologies.

170 Nature-Based Solutions for Urban Renewal in Post-Industrial Cities

3) Through renovation, the underwater forest and water garden of the Moon Lake should reflect the cultural landscape on the shore, which would beautify the environment of the Tianyi pavilion and the Moon Lake. The underwater forest and water garden should strive to become a model of the park landscape and lake management, so that citizens and tourists can enjoy the scenery.

Implementing NBS in the Ningbo Living Lab

Given that Ningbo has already implemented the NBS measures for improving the water quality of the man-made Moon Lake before proGIreg, they are currently only being monitored. Further LL implementation will need to be contextualized within the already-performed implementation (aquatic filtering plants, fry fish, pumps for oxygenizing the water, water filter, new bamboo plantings) and to support the past and ongoing initiatives with three complementary NBS (Table 10.1). Geographical overview of positions for implementing each NBS is depicted in Figure 10.3.

NBS2: Transforming lake sediment into soil fertilizer

This NBS is for reusing lake-bottom sediments and turning waste into treasure. At the bottom of Moon Lake in the LL area, there are many

Table 10.1 Three complementary NBS included in proGIreg for the Living Lab in Ningbo

NBS	Description
2	Utilizing the fertilizer derived from lake sediment into the soil regeneration in a total area of 20 ha green space located in the central district of Ningbo city.
3	Using the emerged macrophytes to renature a 5-km corridor surrounding the urban lake which will limit the runoff from nonpoint pollution sources in urban space.
7	Collecting the integrated dataset of meteorological, hydrological, chemical, and ecological parameters to develop the data-based quantitative protocols and procedures for environmental compensation.

sediments such as sludge. They can release harmful substances to the water body, so the lake must be dredged (Figure 10.4). However, the lake area is huge, and more than 50,000 m^3 of sediments have been removed. Modifying these sediments into soil fertilizers for planting vegetation can greatly protect the environment and save resources. During the time period between January 2019 and December 2020, this NBS was implemented in an area of 20 ha green space surrounding Moon Lake Park. The main beneficiaries of this activity are targeted as the residents living around Moon Lake Park and people who come to the park for tourism.

This project comes from the Moon Lake Water Ecological Comprehensive Improvement Project. Moreover, to cooperate with Tianyi Pavilion and Moon Lake to create a national tourism 5A-level scenic spot, there is an urgent need to improve the water quality of the Moon Lake and beautify Moon Lake Park to adapt it to the requirements of 5A-level scenic spots. The funding for this project comes from two parts: the Ministry of Science and Technology of China and the Ningbo government. The total budget of the implementation is 500,000€. Management structure and responsibilities of this project are organized as in Table 10.2. For implementation, before the sediment is converted into soil fertilizer, the physical and chemical properties of the soil were analyzed first to ensure that the sediment will not cause secondary pollution to the soil. A layer of "ecological phosphorus removal agent" was also laid on the bottom of the lake which can inactivate the phosphorus activity in the water. In this way, not only can the phosphorus

Table 10.2 Management structure and responsibilities for implementing NBS2

Actors	Description
Main partner	The Institute of Urban Environment, Chinese Academy of Sciences (IUE-CAS): organize the transformation of the Moon Lake sediment into soil fertilizer.
2nd Partner	Forestry Bureau of Ningbo City (FBNC): select plant species and use regenerated soil fertilizers for plant cultivation
3rd Partner	Tianhe Aquatic Ecosystem Engineering Co., Ltd.: mainly carry out drainage, dredging, and water storage of lakes
Other stakeholders involved	Ningbo Tianyige Museum and Moon Lake Scenic Area Management Committee

Figure 10.4 Removing sediments in Ningbo Living Lab.

content of the water decrease in a cliff-like manner, but also break the nutritional line of cyanobacteria. However, due to high levels of heavy metals in lake sediments, the implementation of this NBS was terminated.

NBS3: Planting aquatic plants along the shore of the lake

The purpose of planting aquatic plants is to purify water quality, which was successfully experimented in Qinglin Bay Park in Haishu District. It can also contribute to building a green lake shore and benefit the implementation of Ningbo NBS7 (Procedures for environmental compensation activity). As an urban lake located in the urban area, and as a tourist area, there are many hotels and restaurants surrounding the Moon Lake, and the lake water body is often polluted. The water quality of the lake needs to be purified urgently. Planting aquatic plants along the lake can beautify the environment while purifying the water quality. Selected aquatic plants were used to renature a 5-km corridor surrounding the urban Moon Lake Park (Figure 10.6). During the time period between June 2019 and December 2020, this NBS was implemented in an area about 21,641 m^2 (Including 1,918 m^2 of emergent plants and floating plants and 19,723 m^2 of submerged plants) surrounding Moon Lake Park. The main beneficiaries of this activity that are targeted are the schools and residents near Moon Lake Park, as well as citizens of Ningbo city and tourists. The funding for this implementation comes from the same sources as those for the previous NBS implementation, and the total budget of the implementation is also 500,000€. Management structure and responsibilities of this project are organized as in Table 10.3.

Two major aspects were considered for the implementation of this NBS. First, it is needed to choose the type of aquatic plant to ensure that it does not bring the risk of biological invasion. Planted aquatic plants not only need to have a strong purification ability, but also need to be ornamental. Then, rigorous calculations should be carried out to choose the planting location and design the ecological media box (Figure 10.5). The design and layout of the ecological media tank need to be carefully calculated, so that after being beautified, the environment of the Moon Lake provides people with high-quality green space. In addition, an internal circulation system was also installed in the lake to realize the self-purification of the whole lake water every 16 days. The main technologies involved in the implementation include bidirectional living water and purification technology, stepped underwater forest technology, and so on.

The design of planting aquatic plants for the green lake shore requires the joint participation of experts and scholars, local residents, government departments, and enterprises. Experts from the Forestry Bureau of Ningbo City (FBNC) and other research institutions chose species, considering ornamental and

Table 10.3 Management structure and responsibilities for implementing NBS3

Actors	Responsibility
FBNC	Main partner: mainly responsible for the selection and cultivation of aquatic plants.
Ningbo Yilianhuimo Information Technology Co., Ltd.	2nd Partner: responsible for the production of ecological media box mold.
Ningbo Chenyu Construction Engineering Technology Co., Ltd.	3rd Partner: responsible for embankment repair.
Ningbo Tianyige Museum; Moon Lake Scenic Area Management Committee; Work Committee of Moon Lake Street	Other stakeholders involved

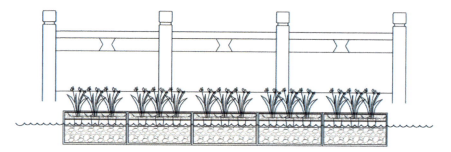

Figure 10.5 A cross-sectional view of ecological media boxes used for planting aquatic plants in Ningbo Living Lab.

Figure 10.6 Planting aquatic plants in Ningbo Living Lab.

purification capabilities and avoiding biological invasion and plant withering. It was also considered necessary to plan the planting of submerged plants and emergent plants. After choosing the species, the ecological media tanks to cultivate aquatic plants were designed by Ningbo Yilianhuimo Information Technology Co., Ltd. Tianhe Aquatic Ecosystem Engineering Co., Ltd. was responsible for calculating the size, location, and spacing of ecological media boxes, while also responsible for plant maintenance and management. Finally, large aquatic plants have been planted along the shore of the Moon Lake to purify water quality and beautify the environment. They enhance the image of Tianyi Pavilion, Moon Lake Scenic Spot and the charm of Ningbo city and provide high-quality green space for surrounding residents.

NBS7: Procedures for environmental compensation

In order to enhance the city's image and promote the development of tourism, Haishu district has gone through long-term investigations and brewing work to do a good job of water quality, sludge, pollutant, and other surveys, and formulate preliminary technical treatment plans. After listening to the opinions of experts in water control, representatives of the National People's Congress and citizens, and optimizing the technical plan for many times, it was decided to start the Moon Lake Water Ecological Comprehensive Improvement Project. This activity is based on the PPP project of Moon Lake Water Ecological Comprehensive Improvement Project. PPP means that in the field of public services, the government adopts a competitive approach to select social capital with investment, operation and management capabilities. The two parties conclude a contract based on the principle of equal negotiation, and the social capital provides public services. The local government of Haishu district pays compensation to the social capital based on the results of public services. The main content of this activity is to evaluate the comprehensive management results of the Moon Lake, that is, collecting meteorological, hydrological, chemical, and ecological data to monitor the environment of the Moon Lake (mainly water environment). If the water quality meets the III level standard, then the government will pay all the costs of the project implementation (Note: In China, water quality is divided into five levels from good to bad: I, II, III, IV, and V).

This NBS was implemented at three observation sites in the Moon Lake (Figure 10.7). The main beneficiaries of this activity are the residents living around Moon Lake Park and tourists of this area. Starting from June 2019, water environment data were collected, but the comprehensive evaluation started at the end of 2021. Funding from the Ministry of Science and Technology supported water quality monitoring, and the Ningbo government has decided to compensate for the cost of the engineering measures. The total budget of the implementation is 50,000€. Management structure and responsibilities of this project is organized as in Table 10.4. In addition, many well-known scientific research institutions and universities such as Wuhan Institute of Aquatic Biology and Ningbo University will carry out scientific research on the management of the Moon Lake's ecological environment and provide scientific theory and technical support for other similar types of lake management in China.

The environmental compensation of the Moon Lake is closely related to NBS2 and NBS3 of Ningbo, because dredging and green lake shore can better affect the water environment of the Moon Lake. However, a major difference of the environmental compensation procedure is that it requires advance payment by social capital. After the results are accepted, the government will compensate the social capital. This process requires the joint participation of enterprises, governments, and third-party supervision agencies. Therefore, before the implementation of the project, experts,

176 Nature-Based Solutions for Urban Renewal in Post-Industrial Cities

Figure 10.7 Three observation sites in the Moon Lake for NBS7.

Table 10.4 Management structure and responsibilities for implementing NBS7

Actors	Responsibility
IUE-CAS	Main partner: organization and coordination of activities, as well as testing of collected water samples.
Tianhe Aquatic Ecosystem Engineering Co., Ltd.	2nd Partner: water samples collection.
Haishu District Government	3rd Partner: provide advice and financial support.
Tianhe Aquatic Ecosystem Engineering Co., Ltd.; Haishu District Government; Ningbo University	Other stakeholders involved

officials, and scholars from various parties conducted detailed discussions, while also consider the wishes of residents. Nearby residents will be affected by the project during the water environment treatment, but will then enjoy the high-quality green space after the water environment treatment. Currently, the environmental compensation has proceeded smoothly. The major achievement of the implementation could be attributed to the introduction of a new type of business model—PPP, which can make full use of social resources for effective project actions.

Transferring knowledge from the Ningbo Living Lab to other contexts

Compared with other Front-Runner Cities (FRC), Ningbo has fewer NBS. Only two official partners were included, but there were also unofficial enterprises and local stakeholders involved in the co-implementation and co-practice. During the co-implementation and co-practice of NBS, the working group was divided into three modules: engineering construction, quality monitoring, and cross-sectional activity. Engineering construction module is about the extraction of lake-bottom sediments and the aquatic vegetation; quality monitoring module is about soil quality monitoring, lake water quality, and plankton monitoring in LL; and cross-sectional activity module is mainly for submission of materials. The purpose of the LL is to obtain certain environmental effects through long-term supervision of the pilot area. In the implementation, we monitored the Ningbo pilot area—the Moon Lake, and obtained the following conclusions:

(1) Weaknesses and challenges in this area: Crowded streets around the Moon Lake; many old residential quarters, old buildings, old streets, and old markets behind the high-rise buildings, with aging infrastructure, insufficient modern facilities and equipment, and many remaining difficulties in coordinating group interests. Some polluted water bodies with seasonal stench have appeared.

(2) Main obstacle in this area: In terms of ecological and environmental restoration, since the Moon Lake is located in the city center, it is difficult and costly to construct some green foundations around, which is the main reason that restricts the further development of the Moon Lake green infrastructure. Ningbo LL implementation was designed to be contextualized within the already-performed implementation (aquatic filtering plants, fry fish, pumps for oxygenizing the water, water filter,

new bamboo plantings) and to support the past and ongoing initiatives, which led to the decision about implementing NBS2, 3 and 7.

(3) Risks in NBS implementation: During the implementation of NBS2 (Transforming lake sediment to soil fertilizer), a fatal risk occurred because the content of heavy metals in lake sediments is too high. If it was converted into fertilizer for planting, low public acceptance will lead to social criticism, thus the implementation of NBS2 was terminated. The improvement of sediments proposed by the involved actors was not enough to explain to the local community that the soil will not be polluted and will not cause harm to people and the environment anymore.

Specifically, the risk identified during NBS implementation reflects at least two shortcomings within the pre-implementation activities: (1) Technical: an in-depth analysis of contaminants in lake sediments should be involved; (2) Co-design: there was a lack of information communication around the involved actors or participants that may be able to identify and remind the risk. Having learned from the risk, Ningbo has introduced new solutions to prevent the highlighted risk: In order to have a more effective soil improvement technology, experts were engaged in soil improvement to give lectures and/or training. In addition, it was decided to facilitate communication between offices through a responsible coordinator who has been managing the exchange of information between experts, staff, and project partners.

References

Xiang, M., Li, Y., Yang, J., Li, Y., Li, F., Hu, B., & Cao, Y. (2020). Assessment of heavy metal pollution in soil and classification of pollution risk management and control zones in the industrial developed city. *Environmental Management*, 66(6), 1105–1119. https://doi.org/10.1007/s00267-020-01370-w

Xu, M., Yang, J., Ren, X., Zhao, H., Gao, F., & Jiang, Y. (2023). Temporal-spatial distribution and health risk assessment of heavy metals in the surface water of Ningbo. *Huanjing Kexue/Environmental Science*, 44(3), 1407–1415. Scopus. https://doi.org/10.13227/j.hjkx.202202195

Evidence-based NBS benefits and related indicators

11

Chiara Baldacchini, Carlo Calfapietra and Martina Ristorini

Introduction

The nature-based solutions (NBS) concept implies the identification of societal challenges (SCs) to be addressed before the implementation, and the verification of the NBS efficacy in connection with them after the implementation, in order to (eventually) change and adapt the intervention, in case of unsatisfactory results. Thus, assessing the benefits produced by an NBS is a crucial step of the implementation process itself, which can be achieved by applying a suitable monitoring and impact evaluation plan. Monitoring is defined as the systematic and standardised gathering of information on a system, using a well-documented approach that can be reliably repeated, so that changes can be compared from time to time and place to place, while impact evaluation focuses on the attribution and causality (Skodra et al., 2021).

Additionally, monitoring is essential to raise awareness and build knowledge about the effectiveness of solutions and provide supporting information which can be used by stakeholders to discuss and select which solutions most closely fit to their aims. Within this context, monitoring and impact evaluation can be considered as important driving factors for evidence-based decision-making for future NBS implementation. It also enables cities to learn from one another by following the best examples and avoiding NBS approaches that are not achieving the target benefits.

DOI: 10.4324/9781003474869-11

This chapter has been made available under a CC-BY license.

180 Nature-Based Solutions for Urban Renewal in Post-Industrial Cities

To obtain a reliable, comprehensive, and holistic assessment of the produced benefits, a well-adapted monitoring and impact evaluation plan should be prepared as a part of the planning/co-design phase and should be regularly updated during the implementation and maintenance periods, as a function of the potential barriers encountered and the newly emerging challenges. The definition of an effective monitoring plan is certainly related to the ability of evaluating the NBS performance or effectiveness, defined as the degree to which targeted problems are solved. Indeed, NBS are implemented to respond to SCs, and it is of utmost importance to verify whether (and to which extent) such SCs have been addressed. Thus, the design of a monitoring and impact evaluation plan is a multi-faceted process that should take into consideration several factors (such as the type of NBS, its extent, the surrounding context, the local SCs, and the potential users) and where general principles are applied for a specific context.

Many tools have already been provided to guide stakeholders into the design of an NBS monitoring and impact evaluation plan. In particular, in recent years, the European Union (EU) has funded a number of Horizon 2020 projects devoted to the implementation and study of NBS interventions, producing among others, two main results: a case study repository, where stakeholders can take advantage of previous experience in planning, implementing, monitoring, and assessing benefits (www.oppla.eu), and a "Handbook for practitioners", where a systematic approach to the building of an NBS monitoring and impact evaluation plan is also described (Skodra et al., 2021). A common framework lies behind this strategy, and it describes the NBS impact in terms of NBS type description, SCs to be addressed, and related KPIs to be assessed. Within this framework, an effective monitoring and impact evaluation plan is described as scientifically sound, practical, and based on a transdisciplinary approach, thus being focused on the integrated evaluation of the provision of cross-sectoral benefits (Skodra et al., 2021). The best compromise should be found between scientific robustness, feasibility of data collection, and data meaningfulness. To reach this goal, good communication among the partners is a prerequisite, to achieve both a resilient monitoring and impact evaluation plan and an efficient data collection.

The process of designing an effective impact monitoring plan may be described by the following eight steps, also sketched in Figure 11.1:

1. **Constructing a theory of change that enables identification of objectives and challenges.** Since the introduction of the NBS concept, many categorisations have been proposed for the SCs that they

are able to address. The most recent proposed classification identifies 12 SC areas: (1) climate resilience, (2) water management, (3) natural and climate hazards, (4) green space management, (5) biodiversity, (6) air quality, (7) place regeneration, (8) knowledge and social capacity building for sustainable urban transformation, (9) participatory planning and governance, (10) social justice and social cohesion, (11) health and well-being, and (12) new economic opportunities and green jobs (Skodra et al., 2021).

2. **Identifying the scales of intervention and the related scales for impact assessment, either spatial or temporal ones.** The specific desired impacts that relate to any of the identified challenges should be defined through spatial mapping and identification of context-based spatial issues. This further leads to the definition of both the scale of NBS intervention and the related spatial scales for impact assessment. Four main spatial scales have been identified and largely described within common assessment frameworks (Leo et al., 2021): element/local/NBS level (building, public space, street), neighbourhood/district/Living Lab level, city level, and regional level. Temporal scales of monitoring are also strongly affected by the NBS intervention size and expected results, as well as by funding duration, if the NBS implementation relates to a specific funding. Three broad timescales have been previously identified for NBS impact evaluation (Raymond et al., 2017): short (within five years), medium (five to ten years), and long term (over ten years). It is worth noting that, while some impacts, such as social or health impacts (e.g., reduction in the prevalence or incidence of different illnesses), require a longer time to become apparent, others, mainly environmental ones, can be verified almost immediately (e.g., the reduction of local temperature through green walls).

3. **Selecting the KPIs that answer the evaluation question(s) and allow the assessment of performance and process.** It is crucial that the KPIs are selected within the framework of a common standard. This will allow further comparison of NBS effectiveness, make results transferable, and thus support decision-makers in evidence-based design of NBS. With this aim, a minimum set of normalised KPIs that may provide a reliable and comprehensive overview of the produced benefits (the so-called "recommended" indicators) has been recently identified and described, for each SC (Wendling et al., 2021).

4. **Identify and collect the data needed to assess the selected indicators.** Specify who is responsible for data collection among the different stakeholders involved and how often data needs to be collected. The desirable quality standards (completeness, precision, uncertainty) should

also be defined and the costs associated with the monitoring estimated. Take into consideration previously collected data (data availability/gap analysis within the local authority and externally), their alignment with selected KPIs, and potential synergies. In case of monitoring activities involving humans, ethical questions of data protection and data management should be taken into consideration. Furthermore, data collection could encounter risks during the monitoring activities. This could be overcome by applying the pre-identified mitigation measures (see step 5) or modifying the monitoring plan, which should be endowed with a certain degree of resilience.

5. **Assess risks associated with data collection activities and mitigation measures.** Risks may arise in data collection activities, such as delays in data collection, low response rate, or unaffordable costs for municipalities. Some of these risks can be identified while preparing the monitoring plan, and the corresponding mitigation measures should be established before starting the data collection (step 4); this will make it easier for local teams to avoid delays and inefficiencies. However, some risks could also occur without being prevented (i.e., the recently occurred COVID-19 pandemic and consequent lockdown). To overcome these difficulties, mitigation measures can also be developed and applied during the data collection, in an iterative process. To allow this, the monitoring plan should include resilience aspects such as not too tight timing or adequate number of replicate samplings.

6. **Implementing the impact evaluation, evaluating positive/negative features of NBS impacts related to the different challenges, analysing and interpreting the findings.** Once data has been identified and collected, the next step is to analyse and interpret it, in order to assess both positive and negative NBS impacts, as well as synergies and trade-offs. If several impacts (positive and/or negative) are considered in relation to an expected objective, the performance evaluation should consider trade-offs and possible differences in timescales over which indicators show that an objective has been achieved or not.

7. **Limitations identification and post-monitoring reassessment of the monitoring plan.** After the impact evaluation, it could become apparent that some monitoring activities have been not as informative as planned and it has not been possible to quantitatively measure certain benefits, either because the selected monitoring tool was poorly designed or due to context-related reasons (e.g., cultural aspects and the reluctance of certain societies to share certain information, or even willingness to participate in certain monitoring activities). The inadequacy of the required data should lead to a reassessment of

the monitoring methodology framework and allow for modification of certain monitoring methods, or even to the establishment of new indicators for future monitoring.

8. **Disseminating results and achieving policy impact.** The wider the dissemination, the more benefits it will produce; citizens will be informed of the activities of their local government, companies will be made aware of business opportunities, and scientists will be able to continue advising on and researching the best methodologies for NBS development and impact assessment. It is important to not only register and report positive results, but also to do so for all the obtained results, in order to help the upscaling of NBS interventions, through the spread of lessons learned.

Although NBS benefit assessment and, thus, the definition of a monitoring plan, is a core driver for NBS decision-making, it often remains a marginal activity in NBS projects. This is likely due to the fact that benefit monitoring and impact evaluation are resource-demanding tasks (also in terms of time and expertise). Nevertheless, it is important that this activity is designed at the early planning phases of an NBS intervention, in order to (1) allocate necessary resources and develop the stakeholder engagement strategy and (2) set up on time an effective pre-implementation (baseline) monitoring.

Baseline data assessment is indeed mandatory, to be used later in the assessment process for the before-and-after comparison. In case it would not be possible to obtain the data required for the baseline assessment, NBS impact and effectiveness can be assessed by comparison with analogous results obtained from control area/site/group, depending on the type of data needed to calculate the selected KPIs. This control area/site/group should be as identical as possible to the area/site/group on which the NBS implementation effectiveness should be evaluated. This means, for instance, that it should be in the same neighbourhood/district/city/region (depending on the scale at which effects are expected) to take local conditions (e.g., climatic conditions and/or cultural ones) into account.

Finally, it is also worth noting that the assessment of NBS effectiveness or impacts is a multi-scale and multi-temporal task. Indicators for urban scales and issues may not be relevant for extra urban context and vice versa. However, if benefit assessment is performed in a rigorous way, the obtained results could be helpful in improving the knowledge about the relation among NBS design and implementation parameters (e.g., scale of intervention, used plant species, implemented co-design and co-maintenance process, etc.) and provided benefits; this being a fundamental step to improve the NBS efficiency in the future, to assess the potential impact of NBS

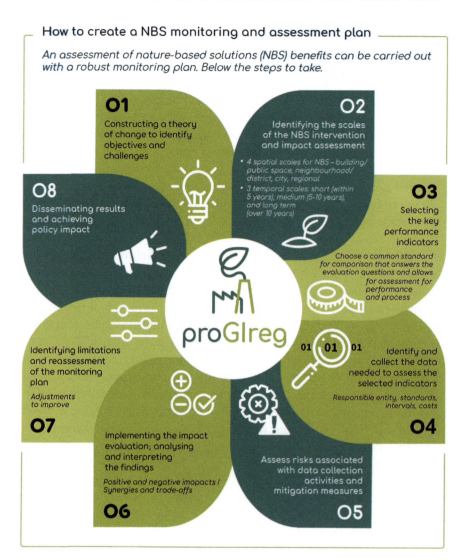

Figure 11.1 Schematic representation of the process of constructing a monitoring and impact evaluation plan (image © ICLEI).

implementation at a larger spatial scale, and to foresee their long-term effects.

The project proGIreg was part of the Horizon 2020 NBS project portfolio, and thus the monitoring activity of the produced benefits was conducted within the EU framework (European Union, 2021). ProGIreg was focused on the implementation of eight different types of NBS, located in post-industrial districts of four cities in Europe and China. Due to the type, the extension, and the spatial distribution of the implemented NBS, their impact was mainly assessed at NBS level (local scale) and Living Lab level (district scale), while data at the city level have been collected only with upscaling purposes (Figure 11.2).

To obtain an overview as comprehensive as possible of the benefits produced by the implemented NBS, the monitoring and impact assessment activities were organised into four assessment domains (Figure 11.3), each one including different SCs of the EU framework. For each domain, several assessment tools were used (and developed) within the project to provide the partners with a qualitative-quantitative description of the produced benefits in terms of the KPIs identified as relevant.

The "Socio-cultural inclusiveness" domain aimed at assessing indicators of socio-psychological benefits and matched SCs 8, 10, and 11 of the EU framework. KPIs such as connectedness to nature, mindfulness, social interaction, and cohesion, and perceived restorativeness were collected by using a general population questionnaire at the district level and an "NBS-visitor questionnaire" tailored for NBS users. Moreover, the Walkability Index, which is an objective measure of how much a particular area is more or less likely to be walkable by people, was calculated at different times. It provides information on the urban structure of a city and districts, and it has also been found to correlate with physical activity of local populations, and with social indicators, such as perceived social interaction.

The "Human health and wellbeing" domain matched SCs 4 and 11 of the EU framework. The collected data provided an evaluation of KPIs on general health, mental health, wellbeing, lifestyle habits, physical activity, and time spent in and perceived quality of the NBS. To be able to detect a change in health and wellbeing indicators that could be attributed to the new NBS, data was collected before and after the NBS implementation. Additionally, the number and demography of visitors and their physical activity levels in the surroundings of the implementation sites were assessed before and after the NBS implementation. Finally, to estimate health benefits of NBS conducted in the context of proGIreg, Health Impact Assessment (HIA) tools were used to estimate the number of cases for different adverse health conditions that could be prevented by the implemented NBS.

The "Ecological and environmental restoration" domain included the SCs 1, 2, 4, 5, and 6 of the EU framework. Ecological and environmental benefits from the proGIreg NBS interventions were estimated experimentally at the NBS scale and, in some cases, upscaled through modelling approaches. For instance, the impact on air quality due to the plant removal from the atmosphere of oxides (CO_2, NO_2), secondary pollutants such as ozone (O_3), and particulate matter (PM) was assessed (Ristorini et al., 2023),

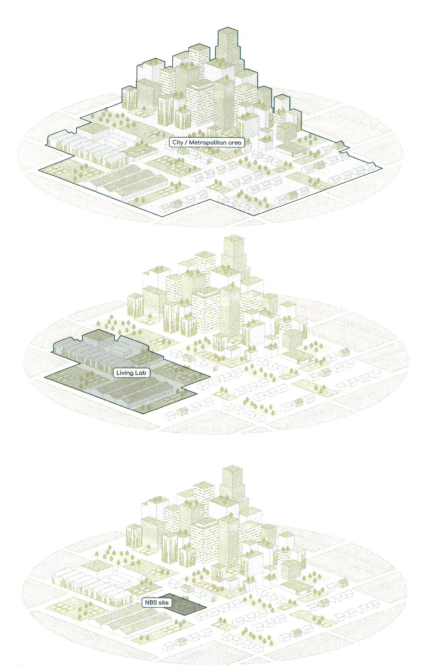

Figure 11.2 Spatial scales of interest in the proGIreg monitoring activity: city (a), Living Lab/district (b), and NBS (c) (image (c) RWTH Institute of Landscape Architecture).

as well as the changes in NBS-level biodiversity of specific target such as pollinators and phytoplankton. The overall environmental footprint of specific implementations was also evaluated in terms of life-cycle assessment (LCA) (Rugani et al., 2024).

The "Economic and labour market benefits" domain matched the SC 12 of the EU framework. Extensive research has shown that increasing NBS in cities is accompanied by multiple direct and indirect economic and labour market benefits. Effects such as increased real estate values, new commercial initiatives, new (and frequently green) job opportunities, and new business opportunities, among others, are all possible when implementing NBS in a city. The main tool applied in proGIreg to capture the direct and indirect economic and labour costs and benefits of the implemented NBS is an "Economic and Labour Market Questionnaire", which was tailored to each combination of NBS + city + stakeholders and administered at least one year after the NBS implementation.

The results obtained in connection with the four domains were analysed in synergistic perspectives, tailoring relevant case studies and newly developed approaches, some of which are presented in detail in the following chapters.

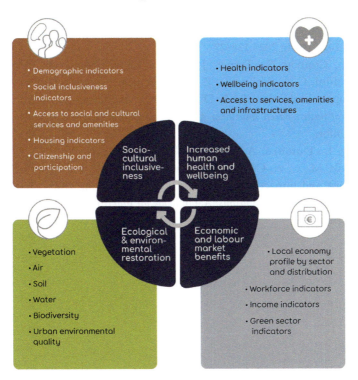

Figure 11.3 ProGIreg assessment domains (image © ICLEI).

References

European Union. (2021). *Evaluating the impact of nature-based solutions: A handbook for practitioners*. European Union.

Leo, L. S., Kalas, M., Baldacchini, C., Budau, O. E., Castellar, J., Comas, J., ... Zavarrone E. (2021). Data requirements. In A. Dumitru & L. Wendling (Eds.), *Evaluating the impact of nature-based solutions: A handbook for practitioners* (pp. 279–369). European Union.

Raymond, C. M., Frantzeskaki, N., Kabisch, N., Berry, P., Breil, M., Niţă, M. R., ... & Calfapietra, C. (2017). A framework for assessing and implementing the co-benefits of nature-based solutions in urban areas. *Environmental Science & Policy, 77*, 15–24. https://doi.org/10.1016/j.envsci.2017.07.008

Ristorini, M., Guidolotti, G., Sgrigna, G., Jafari, M., Knappe, D. R., Garfì, V., ... Calfapietra, C. (2023). Nature-based solutions in post-industrial sites: Integrated evaluation of atmospheric pollution abatement and carbon uptake in a German city. *Urban Climate, 50*, 101579. https://doi.org/10.1016/j.uclim.2023.101579

Rugani, B., Pölling, B., Della Casa, M., Guidolotti, G., Baldacchini, C., Timpe, A., & Calfapietra, C. (2024). Coupled life cycle assessment and business modelling to estimate the sustainability of using regenerated soils in urban forestry as nature-based solutions. *Urban Forestry & Urban Greening, 95*, 128327. https://doi.org/10.1016/j.ufug.2024.128327

Skodra, J., Connop, S., Tacnet, J.-M., Van Cauwenbergh, N., Almassy, D., Baldacchini, C., ... Vojinovic, Z. (2021). Principle guiding NBS performance and impact evaluation. In A. Dumitru & L. Wendling (Eds.), *Evaluating the impact of nature-based solutions: A handbook for practitioners* (pp. 47–70). European Union.

Wendling, L., Dumitru, A., Arnbjerg-Nielsen, K., Baldacchini, C., Connop, S., Dubovik, M., ... zu-Castell Rüdenhausen, M. (2021). Indicators of NBS performance and impact. In A. Dumitru & L. Wendling (Eds.), *Evaluating the impact of nature-based solutions: A handbook for practitioners* (pp. 115–173). European Union.

Benefits from social innovation

12

Egidio Dansero, Luca Battisti, Federico Cuomo, Giacomo Pettenati, Giovanni Sanesi and Giuseppina Spano

Social innovation as a concept has been around for a long time, but it is still debated about its meaning and practice (Seelos and Mair, 2012; Mouleart et al., 2017). In the 1930s, Joseph Schumpeter argued how social innovation was a process that could produce structural changes in society through the activation of the individual or a group of individuals in proposing novel solutions to a problem of collective interest (Schumpeter et al., 2003). Nowadays, the European Commission defines social innovation as '… new ideas that meet social needs, create social relationships and form new collaborations. These innovations can be products, services or models addressing unmet needs more effectively' and that are inherently part of political processes (Ayob et al., 2016).

According to most definitions of social innovation, this concept has a strong spatial dimension that makes it deeply interconnected with environmental characteristics, cultural heritage and history of the place where it occurs (Baker and Mehmood, 2015). In accordance with this perspective, social innovation is considered to be a disruptive process that can radically change citizens' production, mobility, and consumption habits toward more sustainable patterns of living (Angelidou and Psaltoglou, 2017).

Grounding on these assumptions, as stressed by Barbera (2020), the concept of social innovation can take on a twofold connotation: on the one hand, it has gained momentum among policymakers interested in putting in practice alternative policies toward a more fair, sustainable and efficient society; on the other, it constitutes a field of analysis for the social sciences,

DOI: 10.4324/9781003474869-12

This chapter has been made available under a CC-BY license.

with descriptive and explanatory goals, dedicated to a vast array of creative ideas and ways of collaboration aimed at responding to new challenges of contemporary societies.

Often such definitions incorporate in this process the creation of new relationships among multiple actors (Murray et al., 2010; Nicholls and Murdock, 2012; Rhodes et al., 2021). So, the uptake and scaling up of social innovation solutions are often desirable because they can be a source of growth and jobs and can support innovative entrepreneurs and mobilize investors and public organizations. Within proGIreg, social and psychological monitoring protocols will be designed and adopted to monitor social benefits. Social benefits will be monitored in Turin by submitting questionnaires to different group of participants.

State of the art for assessment

The topic of social innovation and nature-based solutions (NBS) is relatively recent, and little research has been conducted that discusses or compares the positive or negative impacts of social innovation. In this regard, Ziegler and colleagues (2022) in their review, highlight how social innovation for biodiversity (which then can be expanded to the topic of NBS) focuses on civic action for changing practices and emphasize how NBS, technology and nature are cross-cutting themes. They also suggest how transformation research needs a focus on social innovation and social exnovation. Likewise, Eichler and Schwarz (2019) pointed out that most of the case studies of social innovation are strongly intertwined with the challenges launched by the 17 Sustainable Development Goals (SDGs) of the United Nations, where NBS might play a pivotal role. According to Galego et al. (2022), social innovation initiatives are able to trigger patterns of collaborative governance by means of participative processes where communities can influence and take part in creative decision-making. In this sense, social innovation is capable of redefining the balance of power between state, market, and society to promote a systemic change in classic patterns of policymaking (Bifulco and Dodaro, 2018).

Nonetheless, both social innovation and NBS present underlying risks when they are not applied in the right way. Some authors highlight the possibility and risk of placing attentions on NBS by focusing more on ecosystem services for which payment might be expected, and that NBS implementation initiatives might not be focused on increasing the level of biodiversity or justice, but from mere discourses on convenience and material needs (Sarkki et al., 2019; Ziegler et al., 2022).

At the same time, there is a need to understand how social innovation can support long-term management and alternative modes of governance of places (Radywyl and Biggs, 2013), while also seeking to protect and privilege the sustaining of places by avoiding a unidimensional transformation (Ahen, 2019; Prasad, 2016; Rivaud and Prevost, 2018).

In this regard, according to recent literature in public policies and urban studies, three main risks might affect social innovations related to environmental and regenerative activities. First, many scholars highlight the potential risk affecting social innovations when they merely consist of symbolic initiatives led by neoliberal models of policymaking, aimed at making disadvantaged communities more productive without actually solving problems of marginalization (Fougère and Meriläinen, 2021). Second, other authors stress the frequent uncertainty and short-term perspective of social innovations related to environmental-preservation activities (Brandsen et al., 2016). In fact, such initiatives are often dependent on temporary projects funded by large-scale institutions. When social innovations are not bridged by consistent and well-structured local public policies, the time limitation risks preventing them from producing long-term benefits for communities and ecosystems. Third, social innovations can be used instrumentally by some local governments interested in cutting public spending in some environmental and social policy areas (Sinclair and Baglioni, 2014). In that case, social innovations proposed from below can be a tool through which to cut public management costs for environmental regeneration or social integration services.

There are various ways to measure the level of Social Innovation, mainly based on surveys of life satisfaction, community benefits generated by businesses, improvements in educational level, and changes in the lifestyle of disadvantaged groups (Mulgan et al., 2013; Unceta et al., 2016). Some authors have proposed assessment models starting from case studies deeply related to environmental issues, such as Secco et al. (2019) who proposed indicators that would consider the level of biodiversity preserved and generated by an innovative agricultural project aimed at rural development.

Regarding the potential of social innovations to transform urban green spaces and cities, a number of articles examine the links between behavioral factors, social innovation, and governance of green spaces. Spijker and Parra (2017) propose research based on document analysis, in-depth interviews, and garden observation, understanding how social-ecological practices have the potential to generate socially innovative physical, social, and political transformations in urban green space governance. At the same time, even if it is not explicitly referred to social innovation, much of the literature on

NBS insists on the need for direct and alternative modes of citizens' and stakeholders' participation to develop solutions that are genuinely innovative and in harmony with surrounding social and environmental systems. According to Ferreira et al. (2020), who have conducted an in-depth textual analysis of 142 papers, the literature on NBSs often refers to terms such as community, participation, policy, and governance that are equally fundamental in defining the concept of social innovation. Citizens' perceptions also play a key role in both concepts of NBSs and social innovation. In the case of natural solutions such as community gardens, hydroponics and aquaponics greenhouses, and pollinator insect gardens, the impact in terms of sense of belonging, new social bounds, and psychological well-being is often assessed by monitoring the perceptions of the citizens involved over time (Campbell-Arvai, 2019). Similarly, social innovation is often seen where there is a change in the perceptions of policymakers related to how to respond to a collective interest challenge by empowering citizens to take part in policies (Angelidou and Psaltoglou, 2017).

Contemplating urban gardens as urban green spaces, but also as potential NBS, the research carried out through in-depth interviews and participant observation by Ulug and Horlings (2018), highlights how it is necessary to define resourcefulness as a process and to highlight the place-based contextual nature of innovative collective (food system) practices.

Despite this, few indicators have been proposed that can assess the link between social innovation and NBSs.

One of the few papers that propose useful indicators for assessing NBS by taking into account the level of social innovation is the paper by Kabisch et al. (2016). The authors paid attention to families of indicators closely related to the social change produced by green solutions, such as those related to psychological health and well-being, the degree of civic participation, or the level of transferability of a given experiment.

However, a larger body of research explores collaborative processes as a way to address or manage factors related to biodiversity loss, highlighting the importance of citizens in such processes (Moore et al., 2014; Spijker and Parra, 2017). In addition to the characterization of citizens and social roles within these processes (Angelidou and Psaltoglou, 2017) and the roles of civil society organizations and networks (Frantzeskaki et al., 2016), it is critical to understand and analyze educational, motivational, and normative determinants of participation in collective action for biodiversity (Tosun and Schoenefeld, 2017; Spijker and Parra, 2017).

NBS, to be well integrated into the urban ecosystem, must be multifunctional and attractive to citizens, who must perceive ecosystem and social benefits (Frantzeskaki, 2019). To achieve these goals, multiple disciplines must be integrated for their design and co-creation (Keesstra et al., 2018).

In general, NBS create new urban green assets and thus often space for new relationships between people and nature and between people in their communities. The proGIreg project, and in general NBS implemented in blighted or post-industrial neighborhoods, has shown that the transformation of the physical characteristics and appearance of public space is associated with the perceived benefits that people assign to it in the form of a sense of place. Such spaces are thus no longer perceived as abandoned or poorly used places to transformed and usable places and community spaces.

Social innovation is thus an important dimension of social capital to consider in the co-creation of NBS (Frantzeskaki, 2019).

Methodology and related indicators

The adoption of qualitative questionnaires over time is considered one of the most promising research methodologies to assess to what extent and how NBS can produce social and psychological benefits. Recently, many empirical studies have adopted this method to understand whether natural solutions can promote socialization, strengthen the sense of belonging, and reduce stress and mental fatigue (Vujcic et al., 2017). This methodology can easily perform comparative studies across different contexts and tailor the complexity of questions to the degree of preparation and characteristics of the respondents (Ferreira et al., 2020). Grounding on these assumptions, some authors put in place questionnaires to specifically deliver case studies' analysis focused on how multifunctional natural spaces could promote children's mental and physical well-being by changing their relational and movement habits (Ramírez-Agudelo et al., 2022). With the outbreak of the COVID-19 pandemic, many authors have pointed out that NBSs can be a valuable remedy to get children back to playing outdoors, learning to care for the surrounding ecosystems and becoming future custodians of urban biodiversity (Grigorescu et al., 2021; Dickson and Gray, 2022). Across Europe and North America, some countries, such as England and Canada, have begun to offer educational programs and spaces for youth activities, such as "Nature Nursery," "Nature Kindergarten," or "Forest Preschool" specifically dedicated to proposing NBS as a learning tool (Jerome at al., 2017; Harwood et al., 2020). Exposure to nature

ProGIreg case studies

The purpose of this case study within the proGIreg project is to monitor any changes in child well-being in terms of perceived psychological restoration and pro-environmental attitude and behavior in schoolchildren during participation in activities near a green wall indoor in a school.

The proGIreg project foresees the preparation and construction of a green wall with dimensions of 20 sq m within a school based in Mirafiori Sud district. It was positioned at a height between 0 and 3 m above the ground floor in a hallway corridor with a large roof window. Site works ended in December 2020, everything was done in one week, during the Christmas holiday, when the school was closed to students. Children from 8 to 12 years old are attending the school in proximity of the green wall.

This study is established within the proGIreg project aiming to set up a series of NBS in the urban environment in order to increase biodiversity, improve citizen well-being, and make the city more sustainable.

To this aim, an NBS-visitor questionnaire, adapted for children, was built to monitor any changes in the aforementioned psychological dimensions. Perceived restorativeness of NBS will be acquired through an adaptation of the version for adults used in the NBS-visitor questionnaire (see Annex 1 and Annex 2). Children are asked to complete the questionnaire under the supervision and support of teachers twice, i.e., for a baseline measure and during a follow-up after approximately two years, in order to evaluate any changes in the aforementioned indicators.

The study lasted about two years and involved teachers and students from the third and fourth grades of the "Gaetano Salvemini" Primary School, Via Negarville, 30/6, 10135- Turin (TO).

Students were asked to fill out a questionnaire before the construction of the green wall in the school and after two years. This questionnaire was aimed at investigating the pro-environmental behaviors enacted by the children and their perceived well-being in relation to carrying out activities in environments that are inside the school but have elements that are reminiscent of natural environments.

Regarding data collection and processing, personal data (contact information on master data such as parents' and child's first and last names) were

Benefits from social innovation 195

Figure 12.1 Green wall in the school based in the Mirafiori Sud district in Turin (image © Monica Rosso (Istituto Comprensivo A. Cairoli).

initially recorded but later anonymized by assigning each child an alpha-numeric code that will allow all data (the parent's answers to the question-naire and the observational data) to be matched. In this way, no one will be able to trace the identity of the participant. The data collected were used for scientific research purposes only. Parents who also provided permission for their child(ren) to participate in the research may withdraw it at any time.

References

Ahen, F. (2019). Making Resource Democracy Radically Meaningful for Stakeowners: Our World, Our Rules? *Sustainability*, 11(19), 23. https://doi.org/10.3390/su11195150.

Angelidou, M., & Psaltoglou, A. (2017). An Empirical Investigation of Social Innovation Initiatives for Sustainable Urban Development. *Sustainable Cities and Society*, 33, 113–125.

Ayob, N., Teasdale, S., & Fagan, K. (2016). How Social Innovation 'Came to Be': Tracing the Evolution of a Contested Concept. *Journal of Social Policy*, 45(4), 635–653.

Baker, S., & Mehmood, A. (2015). Social Innovation and the Governance of Sustainable Places. *Local Environment*, 20(3), 321–334. https://doi.org/10.1080/13549839.2013.842964.

Barbera, F. (2020). L'innovazione Sociale: Aspetti Concettuali, Problematiche Metodologiche E Implicazioni Per L'agenda Della Ricerca. *Polis*, 35(1), 131–148.

Bifulco, L., & Dodaro, M. (2018). Local Welfare Governance and Social Innovation: The Ambivalence of the Political Dimension. In Eraydin, A., Frey, K. (eds.), *Politics and Conflict in Governance and Planning* (pp. 169–185). Routledge.

Brandsen, T., Evers, A., Cattacin, S., & Zimmer, A. (2016). The Good, the Bad and the Ugly in Social Innovation. In: Brandsen, T., Cattacin, S., Evers, A., & Zimmer, A. (eds.), *Social Innovations in the Urban Context. Nonprofit and Civil Society Studies*. Springer. https://doi.org/10.1007/978-3-319-21551-8_25.

Campbell-Arvai, V. (2019). Engaging Urban Nature: Improving Our Understanding of Public Perceptions of the Role of Biodiversity in Cities. *Urban Ecosyst*, 22, 409–423. https://doi.org/10.1007/s11252-018-0821-3.

Dickson, T. J., & Gray, T. L. (2022). Nature-based Solutions: Democratising the Outdoors to be a Vaccine and a Salve for a Neoliberal and COVID-19 Impacted Society. *Journal of Adventure Education and Outdoor Learning*, 22(4), 278–297. https://doi.org/10.1080/14729679.2022.206488

Eichler, G. M., & Schwarz, E. J. (2019). What Sustainable Development Goals Do Social Innovations Address? A Systematic Review and Content Analysis of Social Innovation Literature. *Sustainability*, 11(2), 522. https://doi.org/10.3390/su11020522.

European Commission. Social Innovation. Available at: Social (europa.eu) (accessed July 13, 2022).

Ferreira, V., Barreira, A. P., Loures, L., Antunes, D., & Panagopoulos, T. (2020). Stakeholders' Engagement on Nature-Based Solutions: A Systematic Literature Review. *Sustainability*, 12(2), 640. https://doi.org/10.3390/su12020640.

Fougère, M., & Meriläinen, E. (2021). Exposing Three Dark Sides of Social Innovation through Critical Perspectives on Resilience. *Industry and Innovation*, 28(1), 1–18. https://doi.org/10.1080/13662716.2019.1709420.

Frantzeskaki, N. (2019). Seven lessons for planning nature-based solutions in cities. Environmental Science &Amp; Policy, 93, 101–111. https://doi.org/10.1016/j.envsci.2018.12.033

Frantzeskaki, N. and Kabisch, N. (2016). Designing a knowledge co-production operating space for urban environmental governance—lessons from rotterdam, netherlands and berlin, germany. Environmental Science &Amp; Policy, 62, 90–98. https://doi.org/10.1016/j.envsci.2016.01.010

Frantzeskaki, N., Vandergert, P., Connop, S., Schipper, K., Zwierzchowska, I., Collier, M., & Lodder, M. (2020). Examining the Policy Needs for Implementing Nature-based Solutions in Cities: Findings from City-wide Transdisciplinary Experiences in Glasgow (UK), Genk (Belgium) and Poznań (Poland). *Land Use Policy*, 96, 104688.

Galego, D., Moulaert, F., Brans, M., & Santinha, G. (2022). Social Innovation & Governance: A Scoping Review. *Innovation: The European Journal of Social Science Research*, 35(2), 265–290.

Grigorescu, I., Popovici, E. A., Mocanu, I., Sima, M., Dumitraşcu, M., Damian, N., et al. (2021, September). Urban and Peri-Urban Agriculture as a Nature-Based Solution to Support Food Supply, Health and Well-Being in Bucharest Metropolitan Area During the COVID-19 Pandemic. In *International Conference on Innovation in Urban and Regional Planning* (pp. 29–37). Springer.

Harwood, D., Boileau, E., Dabaja, Z., & Julien, K. (2020). Exploring the National Scope of Outdoor Nature-Based Early Learning Programs in Canada: Findings from a Large-scale Survey Study. *The International Journal of Holistic Early Learning and Development*, 6, 1–24.

Jerome, G., Mell, I., & Shaw, D. (2017). Re-defining the Characteristics of Environmental Volunteering: Creating a Typology of Community-scale Green Infrastructure. *Environmental Research*, 158, 399–408.

Kabisch, N., Strohbach, M., Haase, D., & Kronenberg, J. (2016). Urban Green Space Availability in European Cities. *Ecological Indicators*, 70, 586–596.

Keesstra, S., Nunes, J. P., Novara, A., Finger, D. C., Avelar, D., Kalantari, Z., … & Cerdà, A. (2018). The superior effect of nature based solutions in land management for enhancing ecosystem services. Science of the Total Environment, 610-611, 997–1009. https://doi.org/10.1016/j.scitotenv.2017.08.077

Moore, G., Audrey, S., Barker, M., Bond, L., Bonell, C., Cooper, C., Hardeman, W., Moore, L., O'Cathain, A., Tinati, T., Wight, D., & Baird, J. (2014). Process evaluation in complex public health intervention studies: the need for guidance. Journal of epidemiology and community health, 68(2), 101–102. https://doi.org/10.1136/jech-2013-202869

Mouleart, F., Mehmood, A., MacCallum, D., & Leubolt, B. (2017). *Social Innovation as a Trigger for Transformations: The Role of Research*. European Commission: DG for Research and Innovation.

Mulgan, G., Joseph, K., & Norman, W. (2013). Indicators for social innovation. In F. Gault (Ed.), Handbook of innovation indicators and measurement (pp. 420–438). Edward Elgar Publishing.

Murray, R., Caulier-Grice, J., & Mulgan, G. (2010). National Endowment for Science, Technology and the Arts (Great Britain), and Young Foundation (London, England). The Open Book of Social Innovation, Technology and the Art. National Endowment for Science.

Nicholls, A., & Murdock, A. (2012). *Social Innovation: Blurring Boundaries to Reconfigure Markets*. Palgrave Macmillan.

Prasad, S. C. (2016). Innovating at the Margins: The System of Rice Intensification in India and Transformative Social Innovation. *Ecology and Society*, 21(4), 9.

Radywyl, N., & Biggs, C. (2013). Reclaiming the Commons for Urban Transformation. *Journal of Cleaner Production*, 50, 159–170. https://doi.org/10.1016/j.jclepro.2012.12.020.

Ramírez-Agudelo, N. A., Badia, M., Villares, M., & Roca, E. (2022). Assessing the Benefits of Nature-based Solutions in the Barcelona Metropolitan Area based on Citizen Perceptions. *Nature-Based Solutions*, 2, 100021.

Rhodes, M. L., McQuaid, S., & Donnelly-Cox, G. (2021). Social Innovation and Temporary Innovations Systems (TIS): Insights from Nature-based Solutions in Europe. *Social Enterprise Journal*. https://doi.org/10.1108/SEJ-01-2021-0001.

Rivaud, A., & Prevost, B. (2018). Land Stewardship: An Alternative to Neoliberal Governance for the Conservation of Biodiversity in Natural Areas? *Developpement Durable & Territoires*, 9(3). https://doi.org/10.4000/developpement durable.13051.

Sakhvidi, M. J. Z., Knobel, P., Bauwelinck, M., de Keijzer, C., Boll, L. M., Spano, G., et al. (2022). Greenspace Exposure and Children Behavior: A Systematic Review. *Science of the Total Environment*, 824, 153608. https://doi.org/10.1016/j.scitotenv.2022.153608.

Sarkki, S., Parpan, T., Melnykovych, M., Zahvoyska, L., Derbal, J., Voloshyna, N., & Nijnik, M. (2019). Beyond Participation! Social Innovations Facilitating Movement from Authoritative State to Participatory Forest Governance in Ukraine. *Landscape Ecology*, 34(7), 1601–1618.

Schumpeter, J. A., Becker, M. C., & Knudsen, T. (2003). Entrepreneur. In Roger Koppl (Ed.) *Austrian Economics and Entrepreneurial Studies* (pp. 235–265). Emerald Group Publishing Limited.

Secco, L., Pisani, E., Da Re, R., Rogelja, T., Burlando, C., Vicentini, K., & Nijnjk, M. (2019). Towards a Method of Evaluating Social Innovation in Forest-dependent Rural Communities: First Suggestions from a Science-stakeholder Collaboration. *Forest Policy and Economics*, 104, 9–22.

Seelos, C., & Mair, J. (2012). "Innovation Is Not the Holy Grail", Stanford Social Innovation Review, Fall 2012. Available at: https://ssir.org/articles/entry/innovation_is_not_the_holy_grail# (accessed July 13, 2022).

Sinclair, S., & Baglioni, S. (2014). Social Innovation and Social Policy – Promises and Risks. *Social Policy and Society*, 13(3), 469–476. https://doi.org/10.1017/S1474746414000086.

Spijker, S. N., & Parra, C. (2017). Knitting Green Spaces with the Threads of Social Innovation in Groningen and London. *Journal of Environmental Planning and Management*, 61(5–6), 1011–1032.

Tosun, J. and Schoenefeld, J. J. (2016). Collective climate action and networked climate governance. WIREs Climate Change, 8(1). https://doi.org/10.1002/wcc.440

Ulug, C., & Horlings, L. G. (2018). Connecting Resourcefulness And Social Innovation: Exploring Conditions and Processes in Community Gardens in the Netherlands. *Local Environment*, 24(3), 147–166.

Unceta, A., Castro-Spila, J., & García Fronti, J. (2016). Social Innovation Indicators. *Innovation: The European Journal of Social Science Research*, 29(2), 192–204.

Vujcic, M., Tomicevic-Dubljevic, J., Grbic, M., Lecic-Tosevski, D., Vukovic, O., & Toskovic, O. (2017). Nature Based Solution for Improving Mental Health and Well-being in Urban Areas. *Environmental Research*, 158, 385–392.

Ziegler, R., Balzac-Arroyo, J., Hölsgens, R., Holzgreve, S., Lyon, F., Spangenberg, J. H., & Thapa, P. P. (2022). Social Innovation for Biodiversity: A Literature Review and Research Challenges. *Ecological Economics*, 193, 107336.

200 Nature-Based Solutions for Urban Renewal in Post-Industrial Cities

Appendix: NBS-visitor questionnaire/pre-implementation

Section 1: About You

Participant code (Teacher, please fill in)		_____
1.	How old are you?	_____ years old
2.	Are you male or female?	☐ *Male* ☐ *Female*
3.	What is your class?	_____

Section 2: Environmental Attitudes

4. Please, answer these questions. If you don't understand something, ask for teacher's help!				
		1 False	**2** I don't know	**3** True
a.	If things don't change; we will have a big disaster in the environment soon.	☐	☐	☐
b.	People will someday know enough about how nature works to be able to control it.	☐	☐	☐
c.	When people mess with nature it has bad results.	☐	☐	☐
d.	People are clever enough to keep from ruining the earth.	☐	☐	☐
e.	People are treating nature badly.	☐	☐	☐
f.	I would be willing to go to a school which has a focus on nature.	☐	☐	☐
g.	I believe that artificial light in classrooms should be generated by solar panels.	☐	☐	☐

h.	I would be willing to grow food in the school garden.	☐	☐		☐
i.	I feel more connected with nature when classes are held in outdoor spaces.	☐	☐		☐
j.	It makes me feel better when we have natural day light rather than artificial light all day in classrooms.	☐	☐		☐
k.	People must still obey the laws of nature.	☐	☐		☐
l.	Nature will survive even with our bad habits on earth.	☐	☐		☐
m.	People are supposed to rule over the rest of nature.	☐	☐		☐
n.	Plants and animals have as much right as people to live.	☐	☐		☐

Section 3: Environmental Behavior

5. Please, answer these questions. If you don't understand something, ask for teacher's help!						
		1 Never	**2** Seldom	**3** Sometimes	**4** Usually	**5** Always
a.	I participate in recycling activities at school.	☐	☐	☐	☐	☐
b.	I look at books about the environment (nature, trees, and animals).	☐	☐	☐	☐	☐

c.	I pick up litter left behind by my friends during recess and lunch breaks.	☐	☐	☐	☐	☐
d.	I don't turn on the classroom lights because there is always enough light in my classroom.	☐	☐	☐	☐	☐
e.	I leave the class window open while the heater is working.	☐	☐	☐	☐	☐
f.	I forget to turn off water after washing my hands in the school toilets.	☐	☐	☐	☐	☐
g.	I bring too much food to school and I have to throw away the extra food.	☐	☐	☐	☐	☐
h.	I turn on the air conditioner rather than opening the glass window when it is warm inside.	☐	☐	☐	☐	☐
i.	I forget to turn lights off when I leave a classroom.	☐	☐	☐	☐	☐

Annex 2: NBS-visitor questionnaire/post-implementation

Section 1: About You

Participant code (Teacher, please fill in)	
1.	How old are you?	_____ years old
2.	Are you male or female?	☐ *Male* ☐ *Female*
3.	What is your class?	_____

Section 2: Environmental Attitudes

4. Please, answer these questions. If you don't understand something, ask for teacher's help!			
		1 False	**2** I don't know	**3** True
a.	If things don't change; we will have a big disaster in the environment soon.	☐	☐	☐
b.	People will someday know enough about how nature works to be able to control it.	☐	☐	☐
c.	When people mess with nature it has bad results.	☐	☐	☐
d.	People are clever enough to keep from ruining the earth.	☐	☐	☐
e.	People are treating nature badly.	☐	☐	☐
f.	I would be willing to go to a school which has a focus on nature.	☐	☐	☐
g.	I believe that artificial light in classrooms should be generated by solar panels.	☐	☐	☐

204   Nature-Based Solutions for Urban Renewal in Post-Industrial Cities

h.	I would be willing to grow food in the school garden.	☐	☐	☐
i.	I feel more connected with nature when classes are held in outdoor spaces.	☐	☐	☐
j.	It makes me feel better when we have natural day light rather than artificial light all day in classrooms.	☐	☐	☐
k.	People must still obey the laws of nature.	☐	☐	☐
l.	Nature will survive even with our bad habits on earth.	☐	☐	☐
m.	People are supposed to rule over the rest of nature.	☐	☐	☐
n.	Plants and animals have as much right as people to live.	☐	☐	☐

Section 3: Environmental Behavior

5. Please, answer these questions. If you don't understand something, ask for teacher's help!						
		1 Never	**2** Seldom	**3** Sometimes	**4** Usually	**5** Always
a.	I participate in recycling activities at school or home.	☐	☐	☐	☐	☐
b.	I look at books about the environment (nature, trees, and animals).	☐	☐	☐	☐	☐

c.	I pick up litter left behind by my friends during recess and lunch breaks.	☐	☐	☐	☐	☐
d.	I don't turn on the classroom lights because there is always enough light in my classroom.	☐	☐	☐	☐	☐
e.	I leave the class window open while the heater is working.	☐	☐	☐	☐	☐
f.	I forget to turn off water after washing my hands in the school toilets.	☐	☐	☐	☐	☐
g.	I bring too much food to school and I have to throw away the extra food.	☐	☐	☐	☐	☐
h.	I turn on the air conditioner rather than opening the glass window when it is warm inside.	☐	☐	☐	☐	☐
i.	I forget to turn lights off when I leave a classroom.	☐	☐	☐	☐	☐

206 Nature-Based Solutions for Urban Renewal in Post-Industrial Cities

Section 4: Perceived Restoration Quality of the NBS

In this place I don't think at my worries 👎 👍	In this place everything is just where it should be 👎 👍	This place is interesting 👎 👍	In this place I think about other things, not everyday things 👎 👍
In this place interesting things happen 👎 👍	In this place I am free to play, run and move 👎 👍	In this place I can relax mentally and physically 👎 👍	This place is big enough to be explored 👎 👍
In this place I don't think about things I have to do 👎 👍	This place awakens my curiosity 👎 👍	In this place nobody tells me what to do or think 👎 👍	In this place I only think about things I like 👎 👍
In this place there are lots of things to discover 👎 👍	In this place I don't get bored 👎 👍	I like the room where there is the green wall 👎 👍	

Benefits from circular economy   **13**

Rolf Morgenstern and Bernd Pölling

The European Commission defines nature-based solutions (NBS) as "solutions inspired and supported by nature, designed to address societal challenges which are cost-effective, simultaneously provide environmental, social and economic benefits, and help build resilience" (European Commission, 2016). This definition explicitly includes cost-effectiveness and economic benefits, which can be of direct and indirect nature, including labour benefits. NBS have an economic dimension, which is oftentimes not focused on due to other priorities, like social, cultural, health, ecological, or environmental priorities. Possible economic (co-)benefits include economic and labour market prosperity. Among other things these prosperous features can be increased real estate values in the neighbouring residential areas; higher attractiveness for employees to choose a (close to) NBS business; lower running costs for maintaining areas, like parks and gardens; new commercial initiatives, new (green) job, and employment opportunities as well as new business and start-ups opportunities in or with NBS. Some of these (co-)benefits are especially relevant in urban settings.

Between the framework of NBS and circular economy, overlaps and synergies exist (Pearlmutter et al., 2020). The circular economy approach is to create a closed loop by converting linearity into circularity. "Circular economy aims at the maximum efficiency in the use of finite resources, the wider use of renewable resources, the recovery of materials and products at the end of their useful life, and the regeneration of natural systems" (Stefanakis et al., 2021: 304). It is required to develop the circular model as a wider framework for infrastructure by transitioning from grey to blue and green infrastructure measures.

DOI: 10.4324/9781003474869-13

This chapter has been made available under a CC-BY license.

Circular economy is considered as a way to operationalize and implement the much-discussed concept of sustainable development, which is considered too vague for operationalization and implementation in practice. When talking about circular economy, four "r" occur frequently; reduction, reuse, recycling, and recovery (Kirchherr et al., 2017). However, it is highlighted that many policies focus solely or predominantly on recycling, but neglect or even ignore reduction, reuse, and recovery as core elements of circular economy. Kirchherr et al. (2017) can confirm this in their review paper on circular economy definitions since recycling was found to be named most often. Additionally, they revealed that economic prosperity is the most prominent aim of circular economy, followed by environmental quality.

Methodology and related indicators

Different frameworks have been proposed to analyse and assess the NBS effectiveness from environmental and economic perspectives. For instance, Raymond et al. (2017) put a focus on the assessment of NBS in European urban and peri-urban environments of cities and metropolitan areas. Their work is rooted in the finding that "important questions remain about how to assess the impacts of NBS within and across different societal challenges […] since existing frameworks do not cater for such complexity" (Raymond et al., 2017: 16). The multi-dimensional and multi-functional nature of many NBS challenges the assessment of benefits and co-benefits, including the economic dimension. Additionally, the changing spatial and temporal dynamics have to be taken into account.

For the economic domain, several scholars approached NBS using different methods, including cost-effective assessments, performance assessments against their costs, multi-criteria analysis, and social costs and benefits approach (Pearce et al., 2002; Raymond et al., 2017). Stefanakis et al. (2021) consider NBS as a new tool for economic growth. "Sustainable development requires that environment has a central role in the economic development" (Stefanakis et al., 2021: 308).

The main tool within the proGIreg framework to capture the direct and indirect economic and labour costs and benefits of the NBS implemented is the "Economic and Labour Market Questionnaire" (ELMQ). This questionnaire was administered to the NBS responsible stakeholders only one year after implementation for full overview of costs and estimations of the revenues and maintenance cost structure. In addition to the direct economic costs and benefits, a special focus is laid on the indirect economic values, which are hardly quantifiable and even harder to monetarize. A dilemma exists that NBS are undervalued since you cannot value everything – as highlighted by Edward Barbier, professor in the Department of Economics at Colorado State University, in one of his presentations (naturebasedsolutionsoxford.org).

These aspects were taken into account for setting up a fitting questionnaire to capture information on direct and indirect economic and labour market effects of eight NBS implemented under the proGIreg umbrella. The eight NBS cover a wide set of measures including new soil from deep excavations, community-based urban farming and urban gardening, aquaponics, green walls and roofs, and pollinator biodiversity. The investment and labour intensity differs enormously. Apart from physical implementations, an NBS focusing on strategic planning measures to promote NBS and productive green infrastructure as well as co-design and participatory approaches is followed.

Yet, it has to be mentioned that the selected eight NBS to be concentrated on within the proGIreg project are heterogeneous in many manners, for what reason a fully standardized questionnaire to collect and analyse the direct and indirect economic and labour market questions is not advisable. Nonetheless, the large majority of addressed questions are standardized to allow comparability of findings – given the limited number of NBS implementations mainly in descriptive and qualitative terms.

The standardized questionnaire is comprised of five sections; one introductory section on general aspects and four on costs and benefits of implemented NBS. Two sections are focusing on costs; first on the planning and implementation phase and second on the maintenance/operating phase of the NBS development. The two remaining sections focus on financial revenues and the wider economic impact. The two sections on costs aim for a holistic overview of the financial means required to plan and implement NBS in the various settings of the cities involved in the project. Ideally, these costs pay off after a certain time, although it is not being the primary goal of many NBS implementations. Additionally, it is important to consider that the (financial) benefits are not one by one reaching the main actors responsible for the NBS (or the payer), but indirect financial benefits and remunerations to nearby businesses, real estates, start-ups, individuals, and others have to be seen as a core value of NBS implementations in urban environments.

Costs

The questionnaire asks for splitting the cost items between distinctive phases of NBS developments: planning, implementation, and maintenance. The pioneering character of several preparatory activities to set up NBSs, like new soil from deep excavation sites, aquaponics, or green roofs/walls, turns out to be a costly activity also in the planning phase. This embraces, for instance, (building) permissions, gaining knowledge and expertise, land negotiations, contamination issues on post-industrial sites, and co-design activities. The findings and learnings from the planning phase of NBS developments are on the one hand a significant cost item during the project, but on the other hand allow knowledge transfer and financial savings for replication and upscaling activities carried out by the same stakeholders or others who learn from the proGIreg case studies. Future implementations should financially benefit from the experiences of proGIreg, especially by better-tailored planning. Thus, the questionnaire on the economic and labour market highlights the different phases of NBS developments.

Additionally, the main cost items are collected for systematizing them into categories, like materials, equipment, sub-contracting/tenders for special tasks or expert knowledge, and permission-related costs (including architects, analysis of soil contamination, etc.). One important cost category are labour costs, which are emphasized on in individual questions under the cost sections, both for planning/implementation and maintenance. Labour costs play a significant role for NBS developments. The proGIreg project allows dedicating a huge workload to the different phases of NBS developments. Here, it is also of relevance, how much of the NBS-related labour costs are eligible under proGIreg and how much of the labour costs associated to the NBS are originating from elsewhere. By doing so, it is possible to estimate the immediate role of proGIreg as a lever for job creation. With regard to the post-implementation phase and given the public funding for implementing NBS, the maintenance costs play a crucial role in the longer-term evolution. The responsible key stakeholders, mainly public entities like cities, but also private businesses, cooperatives, associations, or NGOs, aim for financial stability or prosperity over time; for public entities NBS can potentially reduce costs. Here, the maintenance/operating costs are relevant in comparison to the revenue streams and funding options on the other side. Furthermore, the estimation of operating/maintenance costs has to be seen in light of different development stages. Some NBS were already implemented quite a while ago, which allows the indication of details on the maintenance/operating costs. Other NBS developments are at the time of the data collection only shortly accomplished or still in its implementation phase. In these cases, the maintenance/operating-related economic and labour data are only estimations and subject to uncertainties. Additionally, the post-implementation costs might be dynamic and flexible over time – while some costs might be rather static, others increase or decrease with the further development of the implemented NBS. An example for decreasing costs can be achieved by a growing number of green walls and urban gardens. For example, permaculture orchards need some re-planting, watering, etc. at the beginning of the post-implementation phase, but is designed to be largely self-sustaining in the mid- to long-term perspective.

Benefits

While some NBS are implemented with the purpose to create financial revenues, others are not designed for this purpose. However, also these NBS without direct financial revenues for definable entities have the ability to create indirect economic value. Examples are higher rents or real estate values in the neighbourhood, more people using the area and spending money there locally in cafés, shops, etc. Thus, the questionnaire is asking first for quantifiable direct revenue streams and second for estimations of the wider economic impact. The direct financial revenues captured with the NBS can be significant (e.g., social enterprises, NBS start-ups, and cooperation/association), marginal or without direct revenue streams. The questionnaire aims to specify the financial revenues as well as the type of revenue streams exploited. These can be one-off purchase/sales, but also many other types of revenues. Fees or rents for specific goods or services can also create revenues as well as further funding or subsidies, which can (partially) contribute to financial viability of the NBS. Many NBS take advantage of several income paths, e.g., offering paid courses on gardening skills with renting garden parcels or selling of super-local food. Furthermore, financial revenues can also be created by reduced costs. For example, properly managed flower meadows, which are coherently integrated into other landscape activities, e.g., of cities' green departments, have the potential to reduce costs when being implemented on areas with formerly short mowing intervals during vegetation season. The financial revenues captured by the NBS owners or managers might be able to result in financial viable settings, like the rent-a-garden concept, while oftentimes the direct economic dimension alone stays negative, meaning costs more than it benefits the owner or manager of an NBS. Thus, it is pivotal to consider the wider economic impact of NBS as well. These indirect aspects include individuals as well as businesses being situated nearby the NBS or using the NBS implementation as a reference for further applications and tenders. Individuals benefit from NBS via a more attractive environment for outdoor activities promoting health, knowledge creation, and socializing. Furthermore, businesses and owners (of houses, land, etc.) take advantage of rising real estate value. In return, these rising prices have to be mirrored back against possible gentrification issues.

Tailored NBS-specific questions are complementing the standardized part of the questionnaire to take specificities of NBS into account.

Information on economic and labour market issues were not collected from all NBS implementations, but – given the nature of NBS – from a selected number of cases. These are:

- NBS 1 – solar park on Deusenberg site, Dortmund;
- NBS 2 – new soil in Turin;
- NBS 3 – permaculture food forest in Dortmund, Orti Generali in Turin, and the new therapeutic garden in Zagreb;
- NBS 4 – three aquaponics implementations in the three European Front-Runner Cities of Dortmund, Turin, and Zagreb;
- NBS 5 – homeless shelter outdoor green wall in Turin and in Zagreb the NBS4/5 merge of the mini urban farm;
- NBS 7 – strategic planning and environmental compensation protocols in Ningbo, Turin, and Zagreb; and
- NBS 8 – pollinator biodiversity in Dortmund and Turin.

In total, this data set consists of 14 in-depth information on economic and labour market questions. Some insights into case studies are presented in the following section.

ProGIreg case studies

In order to demonstrate economic and labour market benefits originating from NBS developments, this chapter presents three proGIreg case studies – the new therapeutic garden in Zagreb, the rent-a-garden concept of Orti Generali, Turin, and the newly founded association for promoting pollinator biodiversity, Naturfelder e.V. in Dortmund. These case studies are not meant for precisely naming financial numbers for costs and benefits, but to demonstrate the various kinds of costs and benefits, which can occur when implementing NBS on post-industrial sites.

The therapeutic garden in Sesvete, Zagreb, brings together therapy aspects and inclusion for offering a maximum degree of independence and self-realization for the main group of beneficiaries, people with different kinds of disabilities. To do so, the city of Zagreb works together with two day-care centres, which are providing several offers for kids and grown-ups with physical and mental disabilities. To realize the main idea of inclusion, children without any disabilities are another important target group. Although the therapeutic garden does not provide any direct revenue streams, it is important to highlight that the garden potentially generates economic values in the years to come. Among other things, the therapeutic garden can reduce health-related costs for the clients by offering a fulfilling activity outdoors together with people – also out of their comfort zone. Additionally, the beneficiaries can harvest some food and take it home to their families reducing the demand for fresh food purchase. Furthermore, the two day-care centres can benefit from the therapeutic garden since this offer attracts other people with mental or physical disabilities, so that they prefer these two day-care centres to other providers. On the other hand, these very valuable offers, which are providing impact for several beneficiaries, cannot be taken for granted and are not realizable without financial efforts. The planning and implementation of the therapeutic garden costed around 280,000€ and another annual 13,000€ is anticipated for the maintenance. These different kinds of values, which are partly quantifiable, partly not and which are also partly tangible and partly not, cannot be broken down into an easy equation.

Another example with a stronger direct revenue approach is Orti Generali in Turin. The social enterprise in charge (Orti Generali s.r.l. impresa sociale) offers several values on around 5 ha of land, which is leased from the city of Turin: urban farming support service, education social inclusion, community building, and research. The rental of garden plots, a kiosk, and education offers build the core revenue streams of Orti Generali. Their social mission statement becomes obvious when having a closer look to their rental concept. Apart from a so-called "standard gardens", local residents are also able to benefit from the garden when being under 35 years of age or in social difficulties. A certain number of garden plots is rented to these groups with discount rates. The social enterprise aims for economic sustainability within three years of occupation. Their convincing concept allowed them to enlarge their area from 3 to more than 5 ha.

The association "Naturfelder Dortmund e.V." was founded by diverse group of people from Dortmund and surrounding within the course of the co-design phase of NBS 8 "pollinator biodiversity". Their aim is the promotion of flower meadows and other land-use managements within the urban fabric to promote wildlife and biodiversity, especially pollinators. The association is closely collaborating with the cities' green department. This allows the contribution of the wider public since they are able to name public land, which could be transferred from intensively mowed lawns into close to nature areas, including flower meadows. The main revenue stream of "Naturfelder Dortmund e.V." are member fees, but additionally, they are looking for starting crowdfunding campaigns for transferring larger areas, e.g., farmland close to the city, into, for instance, wildlife-friendly flower meadows. The implementation of flower meadows is a rather cheap type of NBS, which has a huge potential for upscaling – meaning that a significant amount of land can be transferred into habitats for wildlife. This includes not only public green areas, but also privately owned or used land, like farmland.

References

European Commission, 2016: Background on green infrastructure. https://ec.europa.eu/environment/nature/ecosystems/background.htm.

Kirchherr, J., D. Reike, M. Hekkert, 2017: Conceptualizing the circular economy: An analysis of 114 definitions, *Resources, Conservation and Recycling*, 127, pp. 221–232.

Naturebasedsolutions.org, 2022: https://www.naturebasedsolutionsinitiative.org/

Pearce, D., D. Moran, D. Biller, 2002: *Handbook of biodiversity valuation: A guide for policy makers*, pp. 1–156, https://doi.org/10.1787/9789264175792-en.

Pearlmutter, D., D. Theochari, T. Nehls, P. Pinho, P. Piro, A. Korolova, S. Papaefthimiou, M.C.G. Mateo, C. Calheiros, I. Zluwa, U. Pitha, P. Schosseler, Y. Florentin, S. Ouannou, E. Gal, A. Aicher, K. Arnold, E. Igondová, B. Pucher, 2020: Enhancing the circular economy with nature-based solutions in the built urban environment: Green building materials, systems and sites, *Blue-Green Systems*, 2, pp. 46–72, https://doi.org/10.2166/bgs.2019.928.

Raymond, C.M., N. Frantzeskaki, N. Kabisch, P. Berry, M. Breil, M.R. Nita, D. Geneletti, C. Calfapietra, 2017: A framework for assessing and implementing the co-benefits of nature-based solutions in urban areas, *Environmental Science Policy*, 77, pp. 15–24, https://doi.org/10.1016/j.envsci.2017.07.008.

Stefanakis, A.I., C. S. Calheiros, I. Nikolaou, 2021: Nature-based solutions as a tool in the new circular economic model for climate change adaptation. *Circular Economy and Sustainability*, 1, pp. 303–318. https://doi.org/10.1007/s43615-021-00022-3.

Benefits for biodiversity 14

*Simona Bonelli, Lingwen Lu,
Marta Depetris, Tian Ruan,
Monica Vercelli, Ruowen Wu
and Yaoyang Xu*

Biodiversity is the sum total of all life forms on earth, including the genetic diversity of organisms, species diversity, and ecosystem diversity, which are formed after a long evolutionary process. Biodiversity is the core component of the earth's life support system, which not only directly provides various foods, medicines, fibres, and building materials but also maintains the living environment necessary for human survival by participating in various biogeochemical cycle processes. Within proGIreg, nature-based solutions (NBS)-tailored biodiversity monitoring protocols have been designed and adopted to monitor biodiversity enhancement. Biodiversity will be monitored in Turin, by studying pollinators, and in Ningbo, focusing on phytoplankton and zooplankton.

State of the art for assessment

NBS are defined by the International Union for Conservation of Nature (IUCN) as "actions to protect, sustainably manage and restore natural or modified ecosystems, that address societal challenges in an effective and adaptive way, simultaneously providing human well-being and benefits for biodiversity".

For these reasons, many NBS are implemented primarily in urban ecosystems, systems altered, and complex designed to primarily provide citizens a range of services, economic and social (Melles, 2005). NBS in an urban environment expands the anthropocentric approach typical of

DOI: 10.4324/9781003474869-14

This chapter has been made available under a CC-BY license.

ecosystem services and uses the collaboration of human beings to enhance the importance of biodiversity in the city. NBS have gained significant attention in recent years as a means of addressing environmental challenges, including the loss of biodiversity. NBS involves the use of natural processes and ecosystems to address problems such as climate change and biodiversity loss. An important aspect of biodiversity is the role of insect pollinators in terrestrial ecosystems of and plankton in aquatic ones. This part provides an overview of the recent research on NBS in connection with biodiversity indicators assessment for insect pollinators and phyto- and zooplankton and the challenges involved in assessing the effectiveness of these solutions in terms of biodiversity enhancement.

Biodiversity indicators for insect pollinators

Insect pollinators, such as bees and butterflies, are essential for maintaining biodiversity in terrestrial ecosystems. They play a critical role in pollinating flowering plants, which provide food and habitat for a wide range of other species. However, insect pollinator populations have been declining in recent years due to factors such as habitat loss, pesticide use, and climate change. Therefore, assessing the effectiveness of NBS in supporting insect pollinators is crucial for biodiversity conservation.

Several indicators have been proposed for assessing the impact of NBS on insect pollinators. These include the abundance and diversity of pollinator species, the abundance and diversity of their host plants, and the connectivity of habitats for pollinator movement. For example, a study by Winfree et al. (2019) assessed the impact of prairie restoration on bee populations and found that restored prairies had higher bee abundance and diversity than non-restored areas.

Biodiversity indicators with phytoplankton and zooplankton

Phytoplankton and zooplankton are important components of aquatic ecosystems, playing a crucial role in nutrient cycling and supporting a wide range of other species. Changes in the abundance and diversity of these organisms can have significant impacts on the structure and function of aquatic ecosystems. Therefore, assessing the effectiveness of NBS in supporting phytoplankton and zooplankton is also important for biodiversity conservation.

Indicators for assessing the impact of NBS on phytoplankton and zooplankton include measures of water quality, nutrient cycling, and community composition. For example, a study by Narayan et al. (2017) assessed the impact of wetland restoration on the abundance and diversity of phytoplankton and zooplankton in the Chesapeake Bay. The authors found that restored wetlands had higher nutrient retention and lower nutrient export than non-restored areas, leading to increased phytoplankton and zooplankton abundance and diversity.

Challenges in assessing the effectiveness of NBS

Assessing the effectiveness of NBS in supporting biodiversity indicators with insect pollinators and phytoplankton and zooplankton can be challenging. One major challenge is the lack of consistent and standardized indicators for measuring these components of biodiversity. Different ecosystems and habitats require different indicators, and it can be difficult to compare results across different studies. Moreover, the effectiveness of NBS in supporting these indicators depends on the scale and context of the intervention.

Another challenge is the potential for unintended consequences or trade-offs associated with NBS. For example, NBS interventions that enhance habitat for one species may have negative impacts on other species or on ecosystem services. Careful planning and stakeholder engagement are crucial for identifying and addressing these trade-offs and ensuring that NBS interventions are socially and environmentally sustainable.

Frameworks in assessing NBS for biodiversity

Several frameworks have been proposed to assess NBS in different contexts. One example is the European Commission's NBS Action Plan, which provides a framework for identifying, assessing, and implementing NBS interventions for biodiversity conservation (European Commission, 2020). The plan emphasizes the importance of integrating NBS into broader landscape and sectoral policies and highlights the need for long-term monitoring and evaluation of NBS interventions to ensure their effectiveness.

Another example is the Biodiversity Indicators for NBS (BINs) framework (IPBES, 2019), which provides a set of indicators for assessing the effectiveness of NBS interventions in supporting biodiversity conservation. The BINs framework includes indicators for both terrestrial and aquatic ecosystems, including indicators for insect pollinators and phytoplankton and zooplankton. The framework emphasizes the importance of context-specific indicators and the need for stakeholder engagement in the assessment process. In 2021, the European Commission proposed

218 Nature-Based Solutions for Urban Renewal in Post-Industrial Cities

a guidebook for practitioners to use to evaluate the effectiveness of NBS. There are 38 indicators that are relevant to enhancing biodiversity and have broad applicability to three kinds of NBS.

Conclusion

In conclusion, NBS have the potential to support biodiversity conservation, including indicators with insect pollinators and phytoplankton and zooplankton. However, assessing the effectiveness of NBS in supporting these ecosystem service can be challenging due to the lack of standardized indicators and the potential for unintended consequences or trade-offs. Frameworks such as the European Commission's NBS Action Plan and the BINs framework provide a useful starting point for assessing NBS interventions and ensuring their effectiveness. Further research and monitoring are needed to improve our understanding of the effectiveness of NBS in supporting biodiversity conservation.

Methodology and related indicators

During proGIreg, the pollinator biodiversity monitoring in Turin and the phytoplankton biodiversity monitoring in Ningbo were carried out using appropriate techniques to obtain key performance indicators (KPIs) suitable for evaluating biodiversity benefits, as follows.

Four of them are connected to Turin and Ningbo's key indicators for assessing the advantages of biodiversity (Table 14.1), according the practitioners' guidebook for evaluating the impact of NBS (European Commission, 2021). The Shannon Diversity Index (9.4) and the Shannon Evenness Index (9.5) are recommended indicators for biodiversity assessment. The Shannon Diversity Index is commonly employed to evaluate species diversity in a specific area, regardless of whether the species present are indigenous, non-indigenous, or invasive. It provides information on the number of different species observed in a defined space and their relative abundance. The Shannon Evenness Index provides information on the relative abundance of each species in a defined area. Extent of habitat for native pollinator species (10.11) refers to the area in a particular ecosystem that can support a plant species' reproduction. The survival and reproduction of these plant species depend on pollinating insects, which need to reside and propagate within the plant's growing area. The size of the habitat range is therefore contingent on the number and distribution of pollinating insects, as well as the habitat characteristics and resources that

Table 14.1 Indicators related to biodiversity enhancement and their general applicability to different types of NBS (modified from European Commission, 2021)

No.	Indicator	Applicability to NBS			Linked methods
		Type 1	Type 2	Type 3	
9.4	Species diversity within a defined area	●	●	●	Located in Ningbo
9.5	Number of species within a defined area	●	●	●	Located in Ningbo
10.11	Extent of habitat for native pollinator species	●	●	●	Located in Turin
10.23	Pollinator species presence	●	●	●	Located in Turin

Notes: Type 1 NBS – minimal or no intervention in ecosystems, with objectives related to maintaining or improving delivery of ecosystem services within and beyond the protected ecosystems.

Type 2 NBS – extensive or intensive management approaches seeking to develop sustainable, multifunctional ecosystems and landscapes in order to improve delivery of ecosystem services relative to conventional interventions.

Type 3 NBS – characterized by highly intensive ecosystem management or creation of new ecosystems.

are necessary for their survival and propagation. Pollinator species presence (10.23) refers to the existence of insect or other small animal species that can pollinate local plants in a specific ecosystem, whether they are native or migratory.

Pollinator biodiversity monitoring

During proGIreg, pollinator monitoring took place according to the European Pollinator Monitoring Scheme (EU-PoMS, Potts et al., 2020). The monitoring consisted in sampling of bees, butterflies, flowers useful for pollinators, and larval food plants, according to the following procedure.

220 Nature-Based Solutions for Urban Renewal in Post-Industrial Cities

Concerning bee surveys, each survey comprises 250 m long linear transects walked in 50 minutes. Each transect start point and direction walked were randomly determined. All unambiguously identifiable bees are recorded and all others that could not be identified in the field are caught with a hand net and retained for later identification. Bee richness and abundance were determined. Insects were identified to species or morphospecies level. Observation sets were made at least one per month, from April to September to cover the main flowering period and bee activity. The observations were conducted between 9:00 A.M. and 5:00 P.M. To monitor butterflies, semi-quantitative surveys were performed by experts walking along fixed-route 300–500 m transects depending on the investigated area (known as "Pollard walk") (Pollard and Yates, 1993). Butterfly species were identified, and individuals of each species counted. Observation sets are made, every two weeks, from April to September to cover the main flowering period and butterfly activity. The observations are conducted between 10:00 A.M. and 3:00 P.M. Windy and rainy days are avoided for all observations and samplings.

Plant surveys, butterfly larval host plants, as well as flower surveys (to identify plants visited by bees and/or butterflies for nectar, pollen, and honeydew) were carried out in parallel to the bee and butterfly surveys along the transects. Plant species were collected and identified, according to Pignatti (2018). The transect walks allow the recording of associations between flowers and bees (essential in studies focusing on pollination ecology and ES) despite the passive sampling methods (i.e., pan traps). Transect walks offer possibilities to evaluate the success of NBS implemented by combining butterfly and bee responses at community level. By sharing monitoring scheme methodologies results are easily comparable. To quantify the biodiversity in a community and the homogeneity of individual distribution between species in the community, Shannon Diversity Index (9.4) and Shannon Evenness Index (9.5) were calculated as KPIs over the monitoring period of both groups. These indices provide valuable information about the fauna richness and composition considering both the number of different species observed and their relative abundances.

Plankton biodiversity monitoring

Traditionally, biodiversity monitoring of phytoplankton and zooplankton in water was conducted by techniques that analyse the water content through microscopy (Munawar et al., 2011) or flow cytometry (Tamm et al., 2018). The latest technology for biodiversity monitoring of phytoplankton and zooplankton in water involves analysing the DNA of phytoplankton and Zooplankton using high-throughput sequencing, a method known as environmental DNA (eDNA) sequencing (Coble et al., 2019). High-throughput sequencing of water samples can detect all the biological species present in the sample, including tiny organisms that cannot be directly observed by traditional methods. In terms of assessing biodiversity, this method can identify various planktonic plant and animal species, providing more comprehensive and accurate biodiversity information (Bista et al., 2017).

In the assessment of the biodiversity of phytoplankton and zooplankton in water during proGIreg, the method of microscope analysis is adopted, since the cost of analysis through microscopy is lower compared to high-throughput sequencing, and since the involved researchers have a previous long-standing technical expertise. The specific analytical steps are as follows:

1. Sample collection: Collect water samples from the lake at different seasons and times. Collect water samples from different locations and depths of the lake to obtain more comprehensive information. Put the water samples into closed containers to ensure they are transported to the laboratory for processing as soon as possible after collection.
2. Sample processing: Filter the water sample to remove impurities, and then use a precipitant to precipitate the plankton. Pour off the supernatant and collect the sediment on microscope slides using transparent tape or a scraper.
3. Microscopic observation: Place the microscope slides on the microscope and observe the plankton in the sediment under magnification. Professional monitors observe their morphology, size, colour, and characteristics to determine different plankton groups and their classifications.
4. Counting and recording: Use a microscope to count their numbers and density and record the classification and quantity of each plankton. These data can be used to calculate various biodiversity indices, such as species richness, evenness, and diversity indices. These indicators can reflect the level of biodiversity of plankton and the impact of environmental changes on them.
5. Data analysis: Collect the data and conduct statistical and comparative analyses of differences in different seasons and times to evaluate the biodiversity level and ecological health status of the lake before and after implementing NBS.

ProGIreg case studies

Case studies in Turin

In Turin, three case studies were analysed as NBS: Gardens in Cascina Piemonte, Green Corridor, and Farfalle in ToUr, which underline their important contribution in maintaining and implementing the level of biodiversity in the city. Specifically, aspects related to pollinators, plant species choice, and Citizen Science have been explored. As a result of the participative approach, biodiversity survey in Cascina Piemonte became the first Italian urban transect to be part of the Citizen Science project of the European Butterfly Monitoring Scheme (eBMS), and it represents the only example of coupled monitoring between butterflies and bees in an urban context, as suggested by Eu-PoMS.

Gardens in Cascina Piemonte (Orti Generali)

Orti Generali was born with the aim of building a model of enterprise for the transformation and management of post-industrial and metropolitan residual agricultural areas based on ecological sustainability and social equity. The implementation of this NBS was concluded in November 2019, in an area of 12.000 sqm. surrounding Cascina Piemonte in Mirafiori Sud district. Monitoring surveys were conducted in 2018, 2019, 2020, and 2021 along two transects called T1 and T2 (Figure 14.1) with different ecological characteristics. The first is characterized by a transitional environment (ecotone) between the river and open grazed meadow; the second one is conducted in the urban gardens. They were managed following organic farming and best practices that consider: the sown of pollinator-friendly plants such as medicinal plants, the consociation techniques intercropping horticultural and ornamental plants, the adoption of ecological infrastructure as pollinator avenue, which acts as an ecological corridor, and improvement of the sources available for the pollinators.

Green Corridor in Turin

The Green Corridor consists of an ecosystem path of about 300 m, capable of redeveloping areas that don't have a strong identity and show climatic criticalities such as the risk of being an "heat island". Thanks to such

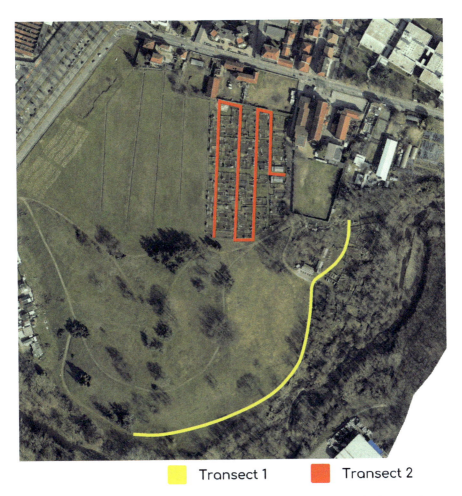

Figure 14.1 Scheme of the transects (T1 and T2) used for butterfly, bee, and flower monitoring for the NBS3.2 in the Cascina Piemonte Park, Mirafiori Sud, Turin. Orthophoto by the City of Turin, 2023.

corridor, pollinating insects will enter urban areas, producing a vital pollination action. Moreover, it will foster processes of involvement, participation, and awareness in the residents. The Green Corridor is developed by incremental area steps inside the Mirafiori district. The area has been identified as part of a series of actions already in place by proGIreg: the development of community farming and beekeeping activities, the organization of crop boxes with the involvement of citizens living in the neighbourhood.

From 2020, a transect walk for monitoring butterfly richness and abundance was defined following the same protocol of the other transects, the

Figure 14.2 Ecotone between the river and open grazed meadow (a) (T1); community gardens (b) and a part of the "pollinator avenue" in community gardens (b) (T2) in Orti Generali in Turin.

fixed route is about 800 m long and totally immersed in an urban environment. This monitoring aims to understand the impact of Green Corridors in the butterfly community and their dispersal abilities in crossing the urban matrix. The green corridor, co-designed and planned for 2020, connects Cascina Piemonte with another green area (Colonnetti Park).

Thanks to the involvement of the Citizen Science project eBMS, European Butterfly Monitoring Scheme, this transect could be monitored by volunteers. The future citizen science data collection complemented the data collected by proGIreg research partners in 2020 and 2021.

Biodiversity and Citizen Science: The Farfalle in Tour project

During proGIreg, butterfly surveys were also carried out by Citizen Scientists' activities in Mental Health Centre gardens close to Cascina Piemonte that contributed to the butterfly survey following the protocols implemented by Farfalle in Tour (http://www.farfalleintour.it/). In brief, users of the

Figure 14.3 Scheme of the "Green corridor" transect for butterfly count in Mirafiori Sud. Orthophoto by the City of Turin, 2023.

Mental Health Centres are directly involved in butterfly monitoring, in collaboration with any other citizen that would join the initiative, always supported by an expert team of University of Turin and educators. Before the surveys begin, the experts must train a group of users about butterfly species morphology (identification) and ecology. This group became the "scientific committee" that during the project transfers the knowledge by teaching to other citizens. According to the extension of the green area, users could monitor butterflies through Pollard walk (described above) or fixed observation point survey, carrying out 15 minutes of observation interspersed with 10 minutes of rest, for 1 hour in total.

Figure 14.4 The Green Corridor in Turin.

A case study in Ningbo

This case study consists in using macrophytes to re-nature a 5 km corridor surrounding the Moon Lake, to limit the runoff from non-point pollution sources in urban space. Aquatic plants used mainly include iris, canna, calamus, and Pontederia, and professional gardeners have planted and maintained them. These plants can not only reduce water pollution but also increase the beauty and ornamentation of Moon Lake Park, which can attract more tourists.

Biodiversity monitoring in the Moon Lake in Ningbo is in charge of IUE-CAS. It involves plankton, which plays an important role in fisheries, water pollution prevention, and environmental impacts of water conservancy projects. Plankton is the primary consumer and producer of freshwater ecosystems and is extremely sensitive to changes in the water environment. Different plankton community structures indicate different water quality conditions. For example, Conochionus and Trichocerca are indicator species for poor nutrient water, and Polyarthra and Bosmina are indicator species for eutrophic water. By investigating the diversity of

Figure 14.5 Moon Lake Park (continuous red line) and three sampling points and the experimenters are collecting plankton samples (image © IUE-CAS).

zooplankton and phytoplankton in the Moon Lake, the impact of NBS3 (using the macrophytes to re-nature a 5 km corridor surrounding the urban lake) can be reflected. The collection of water samples in January 2019, once a week. Three sampling points (S1, S2, S3) have been set up in the Moon Lake, located at the water inlet, the water outlet, and the centre of the lake. All samples were stored in a 4°C refrigerator, until professionals observe the species and quantities of zooplankton and phytoplankton under the microscope.

In March 2020, zooplankton samples collected from January to April 2019 in Moon Lake were analysed by light microscopy. At the genus level, 17 genera of zooplankton have been identified. In January, the relative abundance of Filinia dominates at S1 and S2, while Brachionus dominates at S3. The relative abundance of Brachionus remained high in February and March, and in April the relative abundance of the genera became more balanced.

In May 2020, the types of phytoplankton have also been identified. There are 22 genera of phytoplankton. In general, the composition of phytoplankton in the Moon Lake is quite different in winter and smaller in summer. The relative abundance of Cyclotella is greater in January and February, while Schroederia is greater in June. In July, the relative abundance of Chroomonas, Cyclotella, and Schroederia was large, and the proportion of phytoplankton genera tended to be balanced.

References

Bista, I., Carvalho, G. R., Walsh, K., Seymour, M., Hajibabaei, M., Lallias, D., ... Creer, S. (2017). Annual time-series analysis of aqueous eDNA reveals ecologically relevant dynamics of lake ecosystem biodiversity. *Nature Communications*, 8, 11.

Coble, A. A., Flinders, C. A., Homyack, J. A., Penaluna, B. E., Cronn, R. C. & Weitemier, K. (2019). eDNA as a tool for identifying freshwater species in sustainable forestry: A critical review and potential future applications. *Science of the Total Environment*, 649, 1157–1170.

European Commission. (2020). EU biodiversity strategy for 2030. https://ec.europa.eu/environment/nature/biodiversity/strategy/index_en.htm

European Commission. (2021). Nature-based solutions. https://www.iucn.org/our-work/nature-based-solutions

European Commission, Directorate-General for Research and Innovation. (2021). *Evaluating the impact of nature-based solutions: A handbook for practitioners.* Publications Office of the European Union. https://data.europa.eu/doi/10.2777/244577

IPBES (2019). Global assessment report on biodiversity and ecosystem services of the Intergovernmental Science-Policy Platform on Biodiversity and Ecosystem Services. In E. S. Brondizio, J. Settele, S. Díaz, and H. T. Ngo (editors). IPBES secretariat, Bonn, Germany. 1148 pages. https://doi.org/10.5281/zenodo.3831673

Melles, S. J. (2005). Urban bird diversity as an indicator of human social diversity and economic inequality in Vancouver, British Columbia. *Urban Habitats*, 3(1), 25–48.

Munawar, M., Fitzpatrick, M., Niblock, H. & Lorimer, J. (2011). The relative importance of autotrophic and heterotrophic microbial communities in the planktonic food web of the Bay of Quinte, Lake Ontario 2000–2007. *Aquatic Ecosystem Health & Management*, 14, 21–32.

Narayan, S., Beck, M. W., Wilson, P., Thomas, C. J., Guerrero, A., Shepard, C. C., ... Trespalacios, D. (2017). The value of coastal wetlands for flood damage reduction in the Northeastern USA. *Scientific Reports*, 7, 9463.

Pignatti, S. (2018). *Flora d'Italia, seconda edizione*, Vols. 2–3. Edagricole, Milan.

Pollard, E., & Yates, T. J. (1993). *Monitoring butterflies for ecology and conservation: the British butterfly monitoring scheme.* Springer Science & Business Media.

Potts, S., Dauber, J., Hochkirch, A., Oteman, B., Roy, D., Ahnre, K., ... Vujic, A. (2020). Proposal for an EU Pollinator Monitoring Scheme, EUR 30416 EN, Publications Office of the European Union, Luxembourg. ISBN 978–92-76–23859-1, https://doi.org/10.2760/881843, JRC122225.

Tamm, M., Laas, P., Freiberg, R., Noges, P. & Noges, T. (2018). Parallel assessment of marine autotrophic picoplankton using flow cytometry and chemotaxonomy. *Science of the Total Environment*, 625, 185–193.

Underwood, E., Darwin, G., & Gerritsen, E. (2017). Pollinator initiatives in EU Member States: Success factors and gaps. Report for European Commission under contract for provision of technical support related to Target, 2. Biodiversity Strategy to 2020 – maintaining and restoring ecosystems and their services. ENV.B.2/SER/2016/0018. Institute for European Environmental Policy, Brussels.

Winfree, R., Williams, N. M., Dushoff, J. & Kremen, C. (2019). Native bees provide insurance against ongoing honey bee losses. *Ecology Letters*, 10(11), 1105–1113.

Benefits for health and wellbeing

15

*Mònica Ubalde López,
and Payam Dadvand*

State of the art for assessment

It is projected that by 2050 almost 70% of the global population will be living in urban settings (UN Department of Economic and Social Affairs, 2015). Although cities offer to residents good access to healthcare, education basic services, culture, and are sources of innovation and economic activity (Bettencourt et al., 2007), they also generate multiple environmental hazards. Exposure to high levels of noise, air pollution, heat and low levels of physical activity (PA), and limited access to natural spaces are clear examples (Sallis et al., 2016) of the urban-related environmental and lifestyle determinants that contribute to a high proportion of the population health (Cyril et al., 2013). As natural environments, including green and blue spaces, have been shown to buffer adverse effects of urban settings by improving physical, mental health, and wellbeing (Nieuwenhuijsen et al., 2017), they can contribute with public health practice to maintain and improve population's health following social justice and equity in policies and practice.

Available and increasing evidence have found positive effects of natural outdoor environments on mental health, wellbeing, and quality of life (Gascon et al., 2015; de Keijzer et al., 2016; Spano et al., 2020; Sakhvidi et al., 2022) that vary over different life stages, especially in urban contexts (Browning et al., 2022) (Figure 15.1). Having high residential green areas during pregnancy has been found to be associated with better birth outcomes (e.g., reduced risk for low birth weight and better weight

DOI: 10.4324/9781003474869-15

This chapter has been made available under a CC-BY license.

increase performance) (Akaraci et al., 2020; Zhan et al., 2020), although the relationship between the length of exposure to natural spaces and risk of pre-term birth has shown a less consistent association (Laurent et al., 2013; Hystad et al., 2014; Grazuleviciene et al., 2015; Nichani et al., 2017; Dadvand et al., 2012a, 2012b; Agay-Shay et al., 2014). Evidence is still scarce and has found no significant associations with complications during pregnancy but shows a trend toward a protective association (Zhan et al., 2020). Playing and spending time in natural areas has a crucial role in children's brain development, which is linked to the organic harmony provided by evolutionary bonds between humans and nature (Wilson, 1984; Kellert and Wilson, 1993). Consistent exposure to natural spaces induces positive anatomical changes in the brain (Dadvand et al., 2018), which can enhance cognitive development (e.g., attention and working memory) and improve attention and behavioral disorders (Ricciardi et al., 2022; Markevych et al., 2014; McCormick, 2017). Additionally, it has been observed that children who grow up in green neighborhoods might have better brain development (Dadvand, 2018; Torquati et al., 2017; Bratman et al., 2012), and that students recover better from stress and mental fatigue when they have a landscape view from a school window (Li and Sullivan, 2016). Integrating green areas in school environments has been associated with better academic performance (Ricciardi et al., 2022). Using green spaces benefits social contacts and self-satisfaction among teenagers (Dadvand et al., 2019), facilitates children's interaction and communication, and boosts connection with nature which has been found to promote psychological wellbeing (Shanahan et al., 2015). Among adults, nature exposure has shown to be associated with better cognitive functioning, improved cardiovascular health, and decreased levels of stress (Bratman et al., 2015, Triguero-Mas et al., 2017). Reviews of the available studies found better mental and physical health associated with long-term exposure to green spaces (Gascon et al., 2015; Nieuwenhuijsen et al., 2017). Furthermore, frequent visits to a natural space in the neighborhood are related to improved perceived general health (Gascon et al., 2015), a lower prevalence of depression by promoting positive thoughts (Cox et al., 2017), reduced sense of loneliness, and deceleration of cognitive ageing in the elderly (Ward Thompson et al., 2016; Ricciardi et al., 2022).

There are different mechanisms underlying health and wellbeing benefits of natural spaces by mitigating urban-related hazards: improving environmental quality, enriching biodiversity, increasing PA, stress reduction and attention restoration, and enhancing social interaction and cohesion.

Nature-based solutions (NBS) are solutions that are inspired and supported by nature, which are cost-effective, simultaneously provide environmental, social, and economic benefits, and help build resilience. This natural

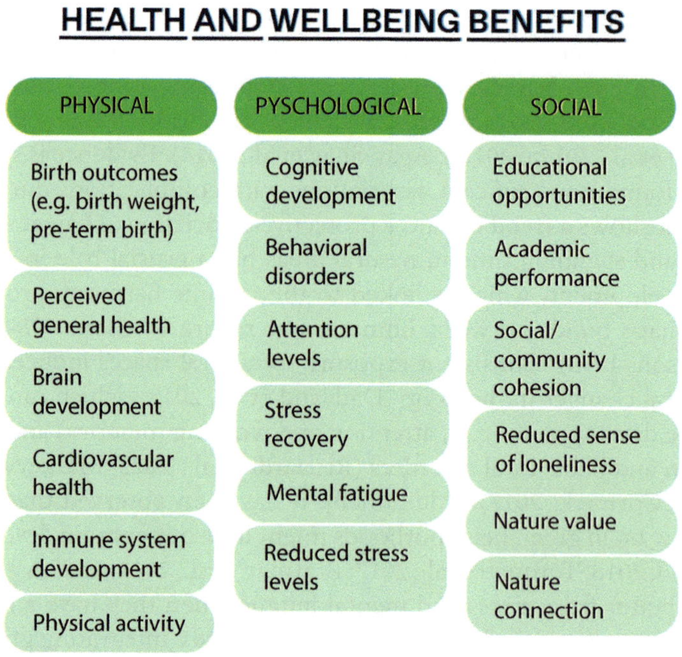

HEALTH AND WELLBEING BENEFITS

PHYSICAL	PYSCHOLOGICAL	SOCIAL
Birth outcomes (e.g. birth weight, pre-term birth)	Cognitive development	Educational opportunities
	Behavioral disorders	Academic performance
Perceived general health	Attention levels	Social/ community cohesion
Brain development	Stress recovery	Reduced sense of loneliness
Cardiovascular health	Mental fatigue	Nature value
Immune system development	Reduced stress levels	Nature connection
Physical activity		

Figure 15.1 Benefits provided to Health and Wellbeing dimensions by access and exposure to green spaces.

infrastructure can service not only to mitigate the effects of the urban heat island effect, aggravated by more recurrent and intense heat waves, but also to improve environmental quality by removing air pollutants and significantly reducing noise levels (Van Renterghem, 2019) by the presence of leaves and woody vegetation (Klingberg et al., 2017). The frequency of visits to natural urban areas could be linked to the thermal comfort and wellbeing perceived by residents during summer and heat waves (Panno et al., 2017), hence to improved health status.

Biodiversity introduced by NBS contributes to buffer adverse human health effects due to the increasing biodiversity loss related to changes in the land use of the rapid urbanization. Biodiversity helps to reduce pathogen transmission and mitigate disease risk (Keesing et al., 2010). Residents health may benefit from increasing permeable soils that are locus of beneficial microbial biodiversity and higher plants and animal species diversity in urban settings (Haaland and van den Bosch, 2015; Gerstenberg and Hofmann, 2016; Marselle et al., 2014). Exposure to biodiversity enhances immune regulation in children by changing gut microbiota and modulating the immune system function (Roslund et al., 2020). More recent evidence has demonstrated that it is plausible that not only psychological alone but also physiological mechanisms exert mental health (Wong & Osborne, 2022) through biodiversity exposure.

Natural infrastructures are capable of reducing perceived stress and restoring attention function, leading to a wide range of health benefits (Nieuwenhuijsen et al., 2017; de Vries et al., 2013; Dadvand et al., 2016). The pleasant characteristics of a natural setting help to reduce negative thoughts and feelings (Ulrich, 1984) and turn to indirect attention (i.e. effortless) (Kaplan & Kaplan, 1989).

In addition, the relationship between greenness and mental health is mediated by the restorative quality of the neighborhood (Dzhambov et al., 2019).

With regard to PA outdoor settings are crucial to promote active behaviors such as walking, cycling, or playing (Wengel and Troelsen, 2020), and hence promote and improve city resident's health. Although few studies have evaluated the mediation effect of natural spaces on PA levels, the current evidence shows heterogeneous results toward a slight mediator role (de Vries et al., 2013; Dadvand et al., 2016). The functional relationship between an individual and the surrounding environment depends on the opportunities for action that the environment provides (Gibson, 1977). The perceived properties of the environment influence the individuals' behavior in that given environment. In this sense, it is most likely that the way the natural areas are planned, designed as well as their quality have an impact and need to be taken into account when assessing its relationship with PA patterns (McCormack et al., 2010). Finally, open natural spaces provide an opportunity to dwellers to interact with each other, increase engagement, affecting positively social cohesion and inclusiveness (Sullivan et al., 2004; Cohen et al., 2008), decreasing loneliness, and enhancing quality of life and life satisfaction (Camps-Calvet et al., 2016; Dzhambov et al., 2018, Maury-Mora et al., 2022; Sia et al., 2022).

Given the many benefits of green spaces for dwellers, cities can be designed healthier and more equitable by implementing natural infrastructures close, available, and of good quality to residents such as green roofs, urban gardens, planting trees, or promoting urban forest. This natural infrastructure should enhance amount and quality and be accessible to all residents to provide an equitable distribution of its potential health and wellbeing benefits (Nieuwenhuijsen et al., 2017; van den Bosch and Nieuwenhuijsen, 2017). However, despite the accumulated evidence on the benefits of natural areas, the knowledge on the public health benefits that implementing new NBS in urban settings (such as providing access to a riverbank or a new park) may provide still deserve a strong interest.

The evaluation of the newly implemented NBS allows estimating the potential health and wellbeing benefits to the residents and guide next steps for the replication and scaling up. Collected data should include indicators on general health, mental health, wellbeing, lifestyle habits, PA, time spent in the new NBS area, perceived quality and satisfaction of/with the NBS, the number and demography of visitors and their PA levels and type.

Methodology and related indicators

To be able to detect a change in health and wellbeing indicators that could be attributed to the new NBS, the most appropriated approach would be to follow a pre-post design with comparable control sites. Data should be collected before and after the NBS implementation in the Living Lab (LL) area. When possible, data will also be collected in a control area (e.g., control street, neighborhood, or district) where no implementation has occurred to be able to attribute the observed changes in health and wellbeing to the new NBS in the implementation areas (LL).

Data on health and wellbeing are collected through items included in three tools administered at two different population levels. At the *city/ Neighborhood level*: (1) the General Questionnaire (GQ), a comprehensive tailored survey on social, health, and economic benefits of new NBS; and at the *NBS level*: (2) the NBS-visitor questionnaire, a survey on perceived social and health benefits derived from a specific NBS; and (3) the SOPARC (System for Observing Play and Recreation in Communities) (McKenzie et al., 2006), a systematic observation tool for recording the characteristics of the users and the types of use in terms of PA levels and type of activities at the specific NBS site. The GQ and the NBS-visitor questionnaire are available in English, Italian, Croatian, and Chinese to be administered in the four Front-Runner Cities (Dortmund, Germany; Turin, Italy; Zagreb, Croatia; and Ningbo, China).

The (1) GQ is conceived for a pre-post interventional study design, starting with a baseline and one or more follow-ups for the assessment of NBS benefits at city or district level. The GQ is based on the adaptation of internationally validated scales and original items created for the purpose. It is designed to be administered through face-to-face interviews to residents of the LL and control districts, before the implementation of the NBS and 24 months after its finalization. The (2) NBS-visitor questionnaire aims to assess social and health benefits obtained from using/visiting the NBS. To evaluate the mid/long-term benefits of the NBS, it is administrated 24 months after the NBS implementation to allow for the consolidation of

Benefits for health and wellbeing **235**

dynamics in the use of new public space. The relevance and originality of this tool lies in the opportunity to monitor different NBS, thus ensuring comparability among multiple NBS types and cities/neighborhoods. The NBS-visitor questionnaire includes items about the perceived health benefits derived from the direct contact with the implemented NBS.

Both questionnaires include items to assess the potential health and wellbeing benefits derived from the NBS that can be grouped in three dimensions: general health, mental health and wellbeing, and PA levels.

General Health is assessed by six indicators: (i) self-perceived health, (ii) bothering symptoms, (iii) somatization, (iv) obesity, (v) smoking, and (vi) alcohol-related habits.

(i) *Self perceived health*: It is assessed by the single-item question of the Short Form of the Self-Reported Health Questionnaire (SF-36) (Ware & Gandek, 1998) which measures the present general health and is widely used as an independent predictor of health outcomes (Desalvo et al., 2006). The question is formulated as "in general, how would you say your health is?" and gives answers to choices structured in a Likert scale as *excellent, very good, good, fair, or poor*. (ii) *Bothering symptoms*: Two items from the Symptom Checklist-90-Revised version (SCL-90r) (Derogatis & Unger, 2010) were included to evaluate the presence of discomforts that the person perceives related to different bodily dysfunctions (cardiovascular, gastro-intestinal, respiratory) (12 bothered symptoms). The answers are rated as *Not at all, A little bit, Moderate, Quite a bit,* and *Extremely*. (iii) *Somatization*: The Four Dimension Symptom Questionnaire (4DSQ) is included to assess self-perceived distress, depression, anxiety, and somatization. It is mainly used in primary general practice with the intention to differentiate between normal distress and psychiatric disorder. For each item, the possible answers are *no, sometimes, regularly, often,* and *very often or constantly*. (iv) *Obesity*: Self-reported height (cm) and weigh (kg) are used to calculate BMI and determine underweight, normal weight, overweight, and obesity. (v) *Smoking* and (vi) *alcohol habits* are assessed by self-reported frequency, former and current consumption, ranking from *never* to *almost daily*.

Mental health and wellbeing are assessed by including items extracted from validated scales that provide five indicators: (i) energy/fatigue, (ii) perceived stress, (iii) anxiety symptoms, (iv) depression, and (v) major life events.

(i) *Energy/fatigue* is measured by four items on energy/fatigue and five items on emotional wellbeing from the mental health domain of the SF-36 (Ware & Gandek, 1998), which aims to measure health status across eight health domains. Possible answers are *all of the time, most of the time, a good bit of the time, some of the time, a little of the time,* and *none of the time*.

These answers are entered as 1 (all of the time) to 6 (none of the time). (ii) *Perceived stress* is assessed by using the Perceived Stress Scale which has four items in only one scale (PSS-4). The PSS-4, developed in 1983, is a widely used method to assess physiological stress perception levels in ten different situations. The four-item scale is the short version of the PSS-10 item scale (Cohen et al., 1983). The questions ask about how often feelings and thoughts bothered during the last month: *never, almost never, fairly often,* and *very often*. Higher scores are correlated to more stress. (iii) *Anxiety symptoms severity* is measured by using the Generalized Anxiety Disorder questionnaire (GAD-7), which is a useful screening tool in primary care and mental health setting for: generalized anxiety, panic, social phobia, and posttraumatic stress disorders. The scale asks for how seven different situations/feelings bothered in the last two weeks rating as *not at all, several days, over half the days,* and *nearly every day*. Higher scores are correlated to more anxiety levels. (iv) *Depression* is measured by the five-item version of the Geriatric Depression Scale (GDS-5) which was created from the 15-item GDS by selecting the items that show the highest correlation with clinical diagnosis of depression, The GDS-5 has shown to be an effective screening test for depression (Hoyl et al., 1999). (v) *Major life events,* which might have affected mental health and wellbeing over the life course, are also assessed by asking whether a major event affected the life in the past four weeks, with answers yes, no, and prefer not to answer.

Physical activity levels: To assess PA levels the short form of the International Physical Activity Questionnaire (IPAQ) was applied. It includes seven questions on frequency and time spent in vigorous, moderate, and mild PA. The answers to these questions will be useful to calculate the amount of energy expenditure carrying out each type of PA in metabolic equivalent of task-min per week (MET-min). The MET-min is the objective measure of the ratio of the rate at which a person expends energy, while performing some specific PA compared to a reference. In short, the MET-min represents the amount of energy expended carrying out PA. A MET is a multiple of the estimated resting energy expenditure, walking is considered to be 3.3 METS, moderate physical activity 4 METS, and vigorous physical activity 8 METS.

(3) The **SOPARC tool** in proGIreg aims to help disentangle the impact of the different NBS by quantifying the number, characteristics (i.e., age group, gender, ethnicity), PA levels, and types of activities of visitors in the surroundings of the implementation sites, before and after NBS implementation. For some of the NBS, the goal is to provide (or provide access to) a space that citizens can use to visit green and/or blue spaces (e.g., providing access to a riverbank, re-naturing a square, etc.) and/or for PA (e.g., bicycle

lane, sport facilities, etc.). SOPARC allows evaluating the effectiveness of NBS on these aspects in terms of the profile of the users and the type and level of PA occurring there or, when possible, on the change in the use of these spaces after their implementation. Data are required to be registered in specific periods of time (morning, lunchtime, afternoon, and evening) and specific days (within one week) by trained fieldworkers. A timely and well-planned training of observers, together with a prior good knowledge of the NBS area, are key to avoid unwanted inter-observer heterogeneity and increase reliability of observations.

The NBS-visitor questionnaire and the SOPARC are the monitoring tools to evaluate health and wellbeing at NBS level. Table 15.1 summarizes NBS-level indicator names and descriptions, assessment domains in proGIreg and, where present, related societal challenge area reflecting the policy priorities of the Europe 2020 strategy (Wendling et al., 2021).

Figure 15.2 Examples of two NBS spots in Dortmund where the SOPARC tool was implemented.

Figure 15.2 (Continued)

Case studies

In four NBS implemented in two of the three Front-Runner Cities, Dortmund (Germany) and Turin (Italy), the monitoring evaluation includes the use of both *NBS-level monitoring* tools (i.e., the NBS visitor questionnaire and SOPARC) to assess the level of perceived quality and health-related self-perceptions of users/visitors at the NBS sites, as well as the changes in the proportion and characteristics of people using the transformed spaces in addition to PA types and levels, respectively.

In the city of Dortmund, the two NBS evaluated transformations include (i) the integration of a sports activity area in an existing park by

Table 15.1 Monitoring tools and related indicators assessed at the NBS-level monitoring, with a short description and indication of the related assessment domain in proGIreg and their reference to the societal challenge area

Monitoring tool	Indicator name	Description	Assessment domain in *proGIreg*	Societal challenge area
NBS-Visitor Questionnaire	Perceived restorativeness of NBS	Perception of restoration coming from an NBS	Sociocultural inclusiveness	11. Health and wellbeing
	Number of and reasons for visits to an NBS area	Visits means discretionary time, ranging from a few minutes out of the home to an all-day trip. Visits may include time spent close to home or further afield, potentially while on holiday	Human health and wellbeing	4. Green space management
	Frequency of use of green and blue spaces	Self-reported time spent in green and blue spaces in hours per week, separately during summer and winter	Human health and wellbeing	4. Green space management
	Satisfaction with green and blue spaces	Self-reported satisfaction with the green and blue spaces in the neighborhood	Human health and wellbeing	4. Green space management
	Self-reported physical activity	Self-reported physical activity in metabolic equivalent of task (MET) minutes per week	Human health and wellbeing	11. Health and wellbeing

(Continued)

Table 15.1 (Continued)

Monitoring tool	Indicator name	Description	Assessment domain in *proGIreg*	Societal challenge area
NBS-Visitor Questionnaire	Self-reported restoration	Restoration Outcome Scale (ROS-S; Subiza-Pérez et al., 2017)	Human health and wellbeing	-
SOPARC	Number and proportion of types of visitors in new recreational areas	The amount and type of people visiting, for leisure purpose over a year, the area where the new infrastructure (both NBS, hybrid solutions and gray infrastructures) is implemented	Human health and wellbeing	4. Green space management
	Observed physical activity levels within NBS	Observed weekly physical activity in the NBS (% over three levels of physical activity [sedentary, walking, or vigorous])	Human health and wellbeing	11. Health and wellbeing

the installation of devices that support physical balance and coordination PA skills, inviting the park users to leave the paths and be active, and (ii) a community-based gardening food forest woodland ecosystem designed for food production. As for the city of Turin, the monitored new NBS consisted of (i) a new forest in a public park and (ii) a community collaborative vegetable garden built in abandoned parts of a public park to encourage community activities such as social farming. The NBS-visitor questionnaire was administrated in the four transformed areas 24 months after the completion of each NBS implementation. To evaluate the changes in the use of the transformed areas with the new NBS implementations, the systematic observation with SOPARC was performed before the transformations and 24 months after its completion.

In addition to the evaluation at the NBS level, the potential benefits on health and wellbeing are also evaluated at the *city/neighborhood level* by administrating the GQ before and 24 months after the completion of the NBS transformations, in both the districts where the NBS are implemented (LL district) and in a control district where no new NBS are implemented.

Finally, indicators that show changes toward benefits in health and wellbeing of residents and users, which could be attributed to the new NBS, will be used to apply Health Impact Assessment (HIA) tools to upscale these results at the city level. HIA is applied to predict health and wellbeing benefits of different "scenarios," for which we can use the input from various stakeholders. The HIA methodology is useful to quantify, for example, the number of cases for different adverse health conditions that could be prevented by NBS. In addition to estimating health and wellbeing benefits of NBS conducted in the context of proGIreg, these tools can also be used to upscale the findings by predicting health benefits of future NBS in the Front-Runner Cities and to replicate them in the proGIreg Follower Cities. This way allows engaging city-to-city exchange to replicative most effective NBS in terms of health and wellbeing attributable benefits.

References

Agay-Shay, K., Peled, A., Crespo, A. V., Peretz, C., Amitai, Y., & Linn, S., et al. (2014). Green spaces and adverse pregnancy outcomes. *Occupational and Environmental Medicine*, 71(8), 562–569.

Akaraci, S., Feng, X., Suesse, T., Jalaludin, B., & Astell-Burt, T. (2020). A systematic review and meta-analysis of associations between green and blue spaces and birth outcomes. *Internatinal Journal of Environmental Research and Public Health*, 17, 2949.

Bettencourt, L.M.A., Lobo, J., Helbing, D., Kühnert, C., & West, G.B. (2007). Growth, innovation, scaling, and the pace of life in cities. *PNAS*, 104(17), 7301–7306.

Bratman, G.N., Hamilton, J.P., & Daily, G.C. (2012). The impacts of nature experience on human cognitive function and mental health. *Annals of the New York Academy of Sciences*, 1249(1), 118–136. https://doi.org/10.1111/j.1749-6632.2011.06400.x.

Bratman, G.N., Hamilton, J.P.,, Hahn, K.S., Daily, G.C., & Gross, J.J. (2015). Nature experience reduces rumination and subgenual prefrontal cortex activation. *Proceedings of the National Academy of Sciences*, 112, 8567–8572.

Browning, M.H., Rigolon, A., & McAnirlin, O. (2022). Where greenspace matters most: A systematic review of urbanicity, greenspace, and physical health. *Landscape and Urban Planning*, 217, 104233.

Camps-Calvet, M., Langemeyer, J., Calvet-Mir, L., & Gómez-Baggethun, E. (2016). Ecosystem services provided by urban gardens in Barcelona, Spain: Insights for policy and planning. *Environmental Science & Policy*, 62, 14–23.

Cohen, D.A., Inagami, S., & Finch, B. (2008). The built environment and collective efficacy. *Health & Place*, 14, 198–208.

Cohen, S., Kamarck, T., & Mermelstein, R. (1983). A global measure of perceived stress. *Journal of Health and Social Behavior*, 24, 386–396.

Cox, D.T., Shanahan, D.F., Hudson, H.L., Plummer, K.E., Siriwardena, G.M., Fuller, R.A., et al. (2017). Doses of neighborhood nature: The benefits for mental health of living with nature. *BioScience*, 67(2), 147–155. https://doi.org/10.1093/biosci/biw173.

Cyril, S., Oldroyd, J.C., & Renzaho, A. (2013). Urbanisation, urbanicity, and health: A systematic review of the reliability and validity of urbanicity scales. *BMC Public Health*, 13, 513.

Dadvand, P., Bartoll, X., Basagaña, X., Dalmau-Bueno, A., Martinez, D., Ambros, A., et al. (2016). *Green spaces and general health: Roles of mental health status, social support, and physical activity. Environment International*, 91, 161–167.

Dadvand, P., de Nazelle, A., Figueras, F., Basagaña, X., Sue, J., Amoly, E., et al. (2012b). Green space, health inequality and pregnancy. *Environment International*, 40, 110–115.

Dadvand, P., de Nazelle, A., Triguero-Mas, M., Schembari, A., Cirach, M., Amoly, E., et al. (2012a). Surrounding greenness and exposure to air pollution during pregnancy: An analysis of personal monitoring data. *Environmental Health Perspectives*, 120(9), 1286–1290.

Dadvand, P., Hariri, S., Abbasi, B., Heshmat, R., Qorbani, M., Motlagh, M. E., et al. (2019). Use of green spaces, self-satisfaction and social contacts in adolescents: A population-based CASPIAN-V study. *Environmental Research*, 168, 171–177.

Dadvand, P., Pujol, J., Macià, D., Martínez-Vilavella, G., Blanco-Hinojo, L., Mortamais, M., et al. (2018). The Association between lifelong green space

Exposure and 3-dimensional brain magnetic resonance imaging in Barcelona schoolchildren. *Environmental Health Perspectives*, 126(2), 027012. https://doi.org/10.1289/EHP1876.

de Keijzer, C., Gascon, M., Nieuwenhuijsen, M.J., & Dadvand, P. (2016). Long-term green space exposure and cognition across the life course: A systematic review. *Current Environmental Health Reports*, 3(4), 468–477.

Derogatis, L.R., & Unger, R. (2010). Symptom checklist-90-revised. *The Corsini encyclopedia of psychology*, 1–2. https://doi.org/10.1002/9780470479216.corpsy0970

DeSalvo, K.B., Bloser, N., Reynolds, K., He, J., & Muntner, P. (2006). Mortality prediction with a single general self-rated health question. *Journal of General Internal Medicine*, 21(3), 267–275.

de Vries, S., van Dillen, S.M.E., Groenewegen, P.P., & Spreeuwenberg, P. (2013). Streetscape greenery and health: Stress, social cohesion and physical activity as mediators. *Social Science & Medicine*, 94, 26–33.

Dzhambov, A.M., Hartig, T., Tilov, B., Atanasova, V., Makakova, D.R., & Dimitrova, D.D. (2019). Residential greenspace is associated with mental health via intertwined capacity-building and capacity-restoring pathways. *Environmental Research*, 178, 108708. https://doi.org/10.1016/j.envres.2019.108708

Dzhambov, A.M., Markevych, I., Hartig, T., Tilov, B., Arabadzhiev, Z., Stoyanov, D., et al. (2018). Multiple pathways link urban green-and bluespace to mental health in young adults. *Environmental Research*, 166, 223–233.

Gascon, M., Triguero-Mas, M., Martínez, D., Dadvand, P., Forns, J., Plasència, A., Nieuwenhuijsen, M.J. (2015). Mental health benefits of long-term exposure to residential green and blue spaces: A systematic review. *International Journal of Environmental Research Public Health*, 12, 4354–4379.

Gerstenberg, T., & Hofmann, M. (2016). Perception and preference of trees: A psychological contribution to tree species selection in urban areas. *Urban Forestry & Urban Greening*, 15, 103–111. https://doi.org/10.1016/j.ufug.2015.12.004.

Gibson, J. J. (1977). The theory of affordances. In R. Shaw & J. Bransford (Eds.), *Perceiving, Acting, and Knowing: Toward an Ecological Psychology* (pp. 67–82). Erlbaum.

Grazuleviciene, R., Danileviciute, A., Dedele, A., Vencloviene, J., Andrusaityte, S., Uždanaviciute, I., et al. (2015). Surrounding greenness, proximity to city parks and pregnancy outcomes in Kaunas cohort study. *International Journal of Hygiene and Environmental Health*, 218(3), 358–365.

Haaland, C., & van den Bosch, C.K. (2015). Challenges and strategies for urban green space planning in cities undergoing densification: A review. *Urban Forestry & Urban Greening*, 14, 760–771. https://doi.org/10.1016/j.ufug.2015.07.009.

Hoyl, M.T., Alessi, C.A., Harker, J.O., Josephson, K.R., Pietruszka, F.M., Koelfgen, M., et al. (1999). Development and testing of a five-item version of the Geriatric Depression Scale. *Journal of the American Geriatrics Society*, 47(7), 873–878. https://doi.org/10.1111/j.1532-5415.1999.tb03848.x.

Hystad, P., Davies, H.W., Frank, L., Van Loon, J., Gehring, U., Tamburic, L., et al. (2014). Residential greenness and birth outcomes: Evaluating the influence of spatially correlated built-environment factors. *Environmental Health Perspectives*, 122(10), 1095–1102.

Kaplan, R., & Kaplan, S. (1989). *The experience of nature: A psychological perspective*. New York: Cambridge University Press.

Keesing, F., Belden, L.K., Daszak, P., Dobson, A., Harvell, C.D., Holt, R.D., et al. (2010). Impacts of biodiversity on the emergence and transmission of infectious diseases. *Nature*, 468, 647–652.

Kellert, S.R., & Wilson, E.O. (1993). *The biophilia hypothesis*. Washington, DC: Island Press.

Klingberg, J., Broberg, M., Strandberg, B., Thorsson, P., & Pleijel, H. (2017). Influence of urban vegetation on air pollution and noise exposure–A case study in Gothenburg, Sweden. *Science of the Total Environment*, 599, 1728–1739.

Laurent, O., Wu, J., Li, L., & Milesi, C. (2013). Green spaces and pregnancy outcomes in Southern California. *Health Place*, 24, 190–195.

Li, D., & Sullivan, W.C. (2016). Impact of views to school landscapes on recovery from stress and mental fatigue. *Landscape and Urban Planning*, 148, 149–158. https://doi.org/10.1016/j.landurbplan.2015.12.015.

Markevych, I., Tiesler, C.M.T., Fuertes, E., Romanos, M., Dadvand, P., Nieuwenhuijsen, M.J., et al. (2014a). Access to urban green spaces and behavioural problems in children: Results from the GINIplus and LISAplus studies. *Environment International*, 71, 29–35.

Marselle, M.R., Irvine, K.N., & Warber, S.L. (2014). Examining group walks in nature and multiple aspects of well-being: A large-scale study. *Ecopsychology*, 6(3), 134–147. https://www.scienceopen.com/document?vid=5f61ac61-ffa3-4b23-a996-8da25e81f4f0

Maury-Mora, M., Gómez-Villarino, M.T., & Varela-Martínez, C. (2022). Urban green spaces and stress during COVID-19 lockdown: A case study for the city of Madrid. *Urban Forestry & Urban Greening*, 69, 127492.

McCormack, G.R., Rock, M., Toohey, A.M., & Hignell, D. (2010). Characteristics of urban parks associated with park use and physical activity: A review of qualitative research. *Health Place*, 16(4), 712–726.

McCormick, R. (2017). Does access to green space impact the mental well-being of children: A systematic review. *Journal of Pediatric Nursing*, 37, 3–7. https://doi.org/10.1016/j.pedn.2017.08.027.

McKenzie, T. L., Cohen, D. A., Sehgal, A., Williamson, S., & Golinelli, D. (2006). System for Observing Play and Recreation in Communities (SOPARC): Reliability and Feasibility Measures. Journal of physical activity & health, 3 Suppl 1, S208–S222.

Nichani, V., Dirks, K., Burns, B., Bird, A., Morton, S., Grant, C. (2017). Green space and pregnancy outcomes: Evidence from growing up in New Zealand. *Health Place*, 46, 21–28.

Nieuwenhuijsen, M.J., Khreis, H., Triguero-Mas, M., Gascon, M., & Dadvand, P. (2017). Fifty shades of green: Pathway to healthy urban living. *Epidemiology*, 28, 63–71.

Panno, A., Carrus, G., Lafortezza, R., Mariani, L., & Sanesi, G. (2017). Nature-based solutions to promote human resilience and wellbeing in cities during increasingly hot summers. *Environmental Research*, 159, 249–256.

Ricciardi, E., Spano, G., Lopez, A., Tinella, L., Clemente, C., Elia, G., et al. (2022). Long-term exposure to greenspace and cognitive function during the lifespan: A systematic review. *International Journal of Environmental Research and Public Health*, 19(18), 11700. https://doi.org/10.3390/ijerph191811700.

Roslund, M.I., Puhakka, R., Grönroos, M., Nurminen, N., Oikarinen, S., Gazali, A.M., et al. (2020). Biodiversity intervention enhances immune regulation and health-associated commensal microbiota among daycare children. *Science Advances*, 6(42), eaba2578.

Sakhvidi, M.J.Z., Knobel, P., Bauwelinck, M., Keijzer, C., Boll, L.M., Spano, G., et al. (2022). Greenspace exposure and children behavior: A systematic review. *Science of the Total Environment*, 824, 153608.

Sallis, J.F., Cerin, E., Conway, T.L., Adams, M.A., Frank, L.D., Pratt, M., et al. (2016). Physical activity in relation to urban environments in 14 cities worldwide: A cross-sectional study. *The Lancet*, 387(10034), 2207–2217. https://doi.org/10.1016/S0140-6736(15)01284-2.

Shanahan, D.F., Fuller, R.A., Bush, R., Lin, B.B., & Gaston, K.J. (2015). the health benefits of urban nature: How much do we need? *BioScience*, 65(5), 476–485. https://doi.org/10.1093/biosci/biv032.

Sia, A., Tan, P.Y., Meng Wong, J.C., Araib, S., Ang, W.F., & Boon, K.H.E. (2022). The impact of gardening on mental resilience in times of stress: A case study during the COVID-19 pandemic in Singapore. *Urban Forestry & Urban Greening*, 68, 127448.

Spano, G., D'Este, M., Giannico, V., Carrus, G., Elia, M., Lafortezza, R., et al. (2020). Are community gardening and horticultural interventions beneficial for psychosocial well-being? A Meta-Analysis. *International Journal of Environmental Research and Public Health*, 17(10), 3584.

Sullivan, W.C., Kuo, F.E., & DePooter, S.F. (2004). The fruit of urban nature: Vital neighbourhood spaces. *Environment and Behavior*, 36(5), 678–700.

Torquati, J., Schutte, A., Kiat, J. (2017). Attentional demands of executive function tasks in indoor and outdoor settings: Behavioral and neuroelectrical evidence. *Children, Youth and Environments*, 27(2), Natural Spaces and Development, pp. 70–92. https://doi.org/10.7721/chilyoutenvi.27.2.0070.

Triguero-Mas, M., Gidlow, C. J., Martínez, D., De Bont, J., Carrasco-Turigas, G., Martínez-Íñiguez, T., et al. (2017). The effect of randomised exposure to different types of natural outdoor environments compared to exposure to an urban environment on people with indications of psychological distress in Catalonia. *PloS one*, 12, e0172200.

Ulrich, R. (1984). View through a window may influence recovery. *Science*, 224(4647), 224–225.

UN Department of Economic and Social Affairs. (2015). *World Urbanization Prospects; The 2014 Revision*. New York: United Nations.

van den Bosch, M., & Nieuwenhuijsen, M. (2017). No time to lose - Green the cities now. *Environment Internatinal*, 99, 343–350.

Van Renterghem, T. (2019). Towards explaining the positive effect of vegetation on the perception of environmental noise. *Urban Forestry & Urban Greening*, 40, 133–144.

Ward Thompson, C., Aspinall, P., Roe, J., Robertson, L., & Miller, D. (2016). Mitigating stress and supporting health in deprived urban communities: The importance of green space and the social environment. *International Journal of Environmental Research and Public Health*, 13(4), 440. https://doi.org/10.3390/ijerph13040440.

Ware, J.E., & Gandek, B. (November 1998). Overview of the SF-36 Health Survey and the International Quality of Life Assessment (IQOLA) Project. *Journal of Clinical Epidemiology*, 51(11), 903–912. https://doi.org/10.1016/s0895-4356(98)00081-x.

Wendling, L., Dumitru, A., Dubovik, M., Laikari, A., zu Castell-Rudenhausen, M., & Fatima, Z. (2021). Indicators of NBS performance and impact. In A. Dumitru, & L. Wendling (Eds.), Evaluating the impact of nature-based solutions: A handbook for practitioners (pp. 115–175). European Commission EC. https://op.europa.eu/s/pajZ

Wengel, T.T.T., & Troelsen, J. (2020). How the urban environment impacts physical activity: A scoping review of the associations between urban planning and physical activity. Copenhagen, The Danish Health Authority. 86 pp.

Wilson, E.O. (1984). *Biophilia*. Cambridge: Harvard University Press.

Wong, Y.S., & Osborne, N.J. (2022). Biodiversity effects on human mental health via microbiota alterations. *International Journal of Environmental Research and Public Health*, 19(19), 11882.

Zhan, Y., Liu, J., Lu, Z., Yue, H., Zhang, J., & Jiang, Y. (2020). Influence of residential greenness on adverse pregnancy outcomes: A systematic review and dose-response meta-analysis. *Science of The Total Environment*, 718, 137420.

Unlocking potential 16

Follower cities' NBS replication strategies for greening urban environments

Margot Olbertz, Codrut Papina, Athina Abatzidi, Melania Blidar, Sandra Dimancescu, Helga Gonçalves, Violeta Irimies, Vasiliki Manaridou, Amra Mehmedić, Teresa Ribeiro, Mirza Sikirić, Bogdan Stanciu and Nerantzia Tzortzi

Introduction of the Follower Cities replication process

ProGIreg has dedicated considerable effort to testing and innovating nature-based solutions (NBS), particularly for rejuvenating post-industrial landscapes and integrating nature into the daily life of local communities in its Front-Runner Cities (FRC) and Follower Cities (FC) alike. Over three years, the four Living Labs (LL) in the FRCs co-created and experimented with NBS pilot interventions, allowing for innovation across a set of eight proGIreg types of NBS. Based on diligent evaluation of FRC experiences and lessons learned from co-designing, co-implementing, and monitoring impacts of NBS interventions, the four FC Cascais, Cluj-Napoca, Piraeus, and Zenica developed context-specific NBS replication strategies.

DOI: 10.4324/9781003474869-16

This chapter has been made available under a CC-BY license.

Creating sustainable urban plans in each FC set out to validate the adaptability and transferability of the proGIreg set of NBS for regenerating urban landscapes and empowering communities. The overarching goal is to enable cities in Europe and worldwide to replicate the proGIreg NBS planning and co-design processes in various local contexts. The NBS implementations and replications serve as valuable knowledge repositories for neighborhood-level regeneration by tailoring NBS to diverse investment priorities, urban landscape conditions, and community needs.

This chapter demonstrates the value of co-designing NBS-driven regeneration strategies for challenging urban environments, considering landscape conditions, urban planning frameworks, green infrastructure (GI) systems, environmental pressures, and community needs. Unlike the FRC, which focused on increasing Technology Readiness Levels (TRL) of eight specific NBS, FC had the opportunity to observe and assess FRC strategic decisions for integrating NBS into their urban frameworks and creating strategies for different time scales. Consequently, the replication process builds on a comprehensive understanding of NBS potentials and pitfalls. However, identifying weak spots in each urban regeneration area (URA) requires an intensive co-design process to develop NBS options adapted to local conditions for remediation. Given a wide array of challenges spanning from low pedestrian accessibility and derelict areas to degrading urban landscapes, FCs addressed these by focusing on converting gray infrastructures, combating heat island effects, improving degraded environmental conditions, and enhancing biodiversity. To achieve this, FCs worked toward reconnecting communities with nature by increasing access to valuable and customized green spaces.

Analyzing how FC applied NBS principles in their strategies can serve as inspiration for other cities kick-starting their green transition. While all NBS adhere to a common formula, factors such as urban morphology, socioeconomic needs, landscape identity, and mobility aspects introduce complexity and variation, aligning with local urban planning frameworks. This complexity enhances the resilience of interventions, establishing multiple connections with present and future resilient urban environments.

To explore the sustainable urban plans developed by each FC, showcasing the innovative use of NBS for urban regeneration interactive maps with so-called project fiches, accessible via the EU website Cordis. The replication process in proGIreg marks a milestone in testing and validating the replication capacities of NBS. Beyond the direct impact and value this work brings to communities and landscapes in FC, it is intended to inspire cities to engage in co-designing NBS strategies and GI. Ultimately, this work aims to guide cities worldwide in collaboratively undertaking urban green transformations using NBS.

Applying the replication methodology in Follower Cities

The FC replication process and the transfer of practices hinge on a newly developed methodology, the "Roadmap toward Urban Planning for NBS" designed to coordinate the co-creation process for developing urban plans in FCs. ProGIreg used this roadmap to conduct the replication process in each FC (Figure 16.1), which is structured as a step-by-step methodology to support the creation of context-specific urban plans.

To effectively transfer and adapt the replication process to each FC context in terms of its greening and urban regeneration strategies, designing the replication strategy drew on extensive literature analysis and previous project work carried out during the NBS co-design and co-implementation processes in FRC. Evidence and knowledge generated in FRC including potential technological and non-technological barriers allowed for integrating key learnings of co-creation processes, pre-empting potential barriers and risks while facilitating knowledge transfer between FRC and FC and external cities. The replication motto "Get inspired, be curious, collaborate, and do not leave any stones unturned" guided the two-year replication process and co-design activities in each FC, leading to a successful co-creation process and is a recommendation for any urban planning process for NBS.

Roadmap and replication toolkit

The roadmap is designed as a flexible tool to fit local needs and NBS development status, considering that any city's starting point and overall conditions are different. It equips cities with tools to guide the development of urban regeneration plans that facilitate the integration of NBS. Creating a Final Urban Plans that is integrated into the local planning framework is the key output of this process.

The methodology is composed of three phases, broken down into distinct blocks, entailing a series of interrelated steps. Elaborating each roadmap phase, block, and step facilitated the understanding of relationships between steps and planned activities while identifying key requirements for a successful replication process. The logic of correlating the steps is meant to inspire cities to engage in their own co-design regeneration strategies. The process is not linear but needs to be flexible to allow for reevaluation and adjustment to avoid risks and adapt to local frameworks and changing scenarios. Throughout the roadmap there are several rerouting points, which indicate the need to check and validate previous steps. Key milestones indicate the point when relevant decision has been already made. Each step builds upon the previous one, ensuring a consolidated wealth of knowledge for diverse contexts while permitting structural changes if needed. This adaptability ensures that cities can tailor the roadmap to meet local needs and new conditions.

Given the importance of empowering citizens to shape their living conditions, the process is highly participatory, emphasizing stakeholder involvement, and advocating multi-leveled collaboration. Its multi-directional nature supports several implementation paths through interaction among actors, experts, and the municipality, facilitated by a scenario-building approach. Cities should plan NBS integration across three different time frames: 5 years for regeneration actions and

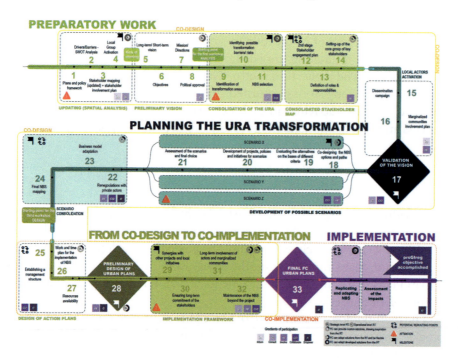

Figure 16.1 Roadmap toward urban plans for NBS in FC (URBASOFIA).

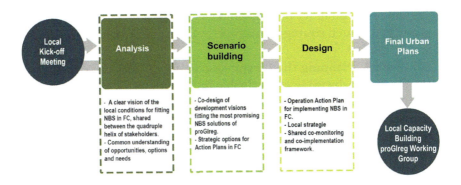

Figure 16.2 Main co-design workshops, integrated in the roadmap process (URBASOFIA).

immediate actions (focused more on pilots and simple investments), 10 years for intermediate transitions (including complex projects and strategic investments), and 15 years for complex transformations (including policy measures adaptations, regulatory measures, and projects with long-term impact).

Integrated in the roadmap replication process are three major co-creation workshops, which are key points in validating the multi-leveled analysis and research activities with the main objectives—decision-making and collaboration (Figure 16.2). The local kick-off and local actors network activation is coupled with the analysis workshop to guide developing a vision for the URA and identifying general barriers and risks. To maintain momentum in the process, FC engaged in a series of bilateral activities: individual interviews, discussions with the residents, and large-scale questionnaire and surveys. Based on this, co-designing scenarios lead to the final design phase, in which specific investments and NBS sites are detailed in the project fiches. To conclude, the Final Urban Plans should be discussed/debated in the final local capacity building event.

To provide an overview, the three roadmap phases can be characterized as follows:

Phase 1: Preparatory work. This phase must be explorative, e.g., examining URA conditions and community needs in detail, identifying key intervention areas, barriers, and opportunities. It seeks to establish a common and shared vision among stakeholders, to explore synergies between the NBS strategy and other planned investments, to understand the conditions and limits of urban planning frameworks, and to define roles and responsibilities of stakeholders. The goal of Phase 1 is to deliver a vision and related strategic objectives, supported by political leaders and decision-makers.

In Phase 2: **Planning the LL transformation**. NBS interventions in the URA focus on achievable and valuable interventions. At this stage, the set of NBS is finalized, and potential implementation models are defined. The scenarios developed are strategic narratives that consider various and plausible combinations of local resource mobilization, based on previously identified trends, underpinning the URA vision.

Phase 3: From co-design to co-implementation. It lays the foundation for strategy implementation. Cities create project fiches, detailing investment procedures, contracting work, and the development of technical and architectural plans. Using project fiches is a good option for cities when the exact funding mechanism or investment framework is not yet decided. These fiches serve as a testament to the needs of local communities and outline how to address them while respecting key NBS criteria.

Figure 16.3 Example of project fiche, page 1 and page 3 of C6.2+8.2 Pollinator-friendly pedestrian trail along the river (URBASOFIA, CASCAIS AMBIENTE).

Unlocking potential 253

Figure 16.4 Example of project fiche, page 1 and page 5 of C6.3 Green corridors in former industrial areas (URBASOFIA).

Key output—FC Final Urban Plans

The FC urban plans are based on the outcomes of the roadmap phases, comprising a comprehensive analysis of the URA, site selection, socioeconomic drivers, and local development priorities across short-, medium-, and long-term perspectives, defining prerequisites, and key factors required for each NBS intervention. The Final Urban Plans for NBS are strongly aligned with local planning and strategic frameworks. For instance, developing the urban plan in FC Zenica coincided with the finalization of the city's Development Strategy 2023–2027; therefore, proGIreg results could be fully incorporated. The example shows that integration into strategic planning documents depends on the timing of key local strategic processes and decision-making and political support.

The Final Urban Plan is a map displaying the URA's strategic locations, types of NBS selected, GI system, and natural landscapes. It represents the spatial component of the strategy created. Alongside, the project fiches for each NBS interventions proposed serve decision-makers and communities on how to design, pilot, and sustain the planned NBS. Depending on the specific requirements there are multiple types of proposals (fiches).

254 Nature-Based Solutions for Urban Renewal in Post-Industrial Cities

All documents can be found on the EU project website: https://cordis.europa.eu/project/id/776528/results

- Simple investments in NBS interventions are easily implementable, drawing from past municipal experiences and requiring minimal resources and know-how. While not necessarily cheaper, they are generally considered affordable and can be realized quickly by public authorities in collaboration with local actors. See project fiche on NBS 6.2 for FC Cascais.
- Complex projects are spatially defined interventions requiring longer time frames, a wider range of stakeholders, and extensive expertise for implementation. Although they tend to be more expensive, they are strategic investments at the city or district level. See project fiche for NBS 6.2 for FC Zenica.
- Policy and zoning regulation project fiches propose changes to local frameworks to facilitate broader NBS adaptation. They aim to create conducive environments at urbanistic, administrative, and legislative levels for NBS interventions on a larger scale and across multiple sites. See project fiches for NBS 3.1 and 8.1 in FC Piraeus.
- The urbanistic program comprises a collection of smaller-scale interventions that align with the same objectives and design requirements. Detailed geographical areas of influence are outlined in project fiches, which can encompass entire neighborhoods or specific sites identified through the co-creation process. See project fiche NBS 6.2 for FC Cluj-Napoca.

Key challenges and learnings of the FC replication process

The true value of the proGIreg approach lies in the diverse range of adaptable NBS to various conditions. Consequently, the roadmap can be universally used for planning other NBS types beyond the proGIreg NBS.

Figure 16.3 shows that 33 NBS are planned across the four URAs across the different NBS types. The distribution highlights two NBS that are most prominently featured in FC strategies: NBS 3 Community-based urban gardens and farms and NBS 6 Green Corridors.

Two solutions have not been chosen by any FC, namely, NBS 2 New Regenerated Soil and NBS 4 Aquaponics, primarily due to the specific stakeholder setups and access to resource flows they require.

FCs developed closely connected stakeholder networks. This collaboration with local communities and stakeholders has been instrumental in

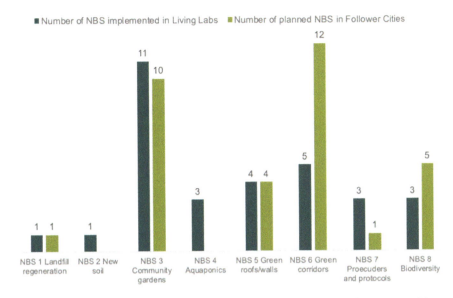

Figure 16.5 Overview of NBS implemented in FRC and planned in FC (RWTH).

driving the urban plans forward, ensuring relevance and fostering a sense of ownership among residents while aiming at setting new local standards for community-oriented green facilities.

To facilitate the widespread use of the replication strategy and use of the roadmap, the following recommendations on key steps and key lessons learned are summarized below:

- The process of site identification takes time. Allocate sufficient resources for the identification of strategic intervention sites. Mapping private and public plots is a requirement. Collaborate with local stakeholders and communities to understand their views on potential NBS sites and their relation to day-to-day activities in these areas.
- Use the scenario process to discuss different design options and their potential local impact, and different operation models with citizens and future users, allowing for collaboratively deciding what fits best to choose the most plausible scenario for site design.
- Ensure a smooth transition from co-design to co-implementation by clearly determining the roles and responsibilities of involved parties, maintaining stakeholder engagement, and collaborating with local authorities for the implementation and subsequent management of the NBS intervention.

256 Nature-Based Solutions for Urban Renewal in Post-Industrial Cities

- Develop and refine business models to prepare for and sustain the NBS intervention.
- Finally, it needs to be stressed that considering NBS adaptation and integration as a fundamental and valuable change compared to conventional GI management practices requires shifting local mindsets and urban planning practices. This shift involves addressing local environmental challenges and social issues through co-creation and co-ownership within multi-stakeholder constellations, as opposed to traditional top-down planning and implementation processes.

Critical reflection of the roadmap's usefulness

Outside the context of applied research projects, such intensive co-creation efforts for NBS might seem too resource intense. However, the experiences of the four FCs highlight the fact that without the incremental, iterative co-design-based process, finding most suitable sites for NBS would have proved difficult, unless compromising on the likelihood of community NBS ownership. In addition, it helps consolidate a citizen-driven new, green, nature-inclusive identity for the URA.

Tested over a two-year period, the roadmap has proved a valuable methodological tool for coordinating the co-creation processes to develop urban plans for NBS, providing guidance in collaborative efforts for urban green transformations. All FCs successfully implemented the Roadmap's methodological steps such as creating the vision for the URA, developing scenarios (do-it-all, do-something-meaningful, business-as-usual), which culminated in the spatial representation of regeneration strategies. The roadmap key principles proved viable for incrementally planning co-design activities for different local settings with the goal of achieving effective and targeted NBS strategies. This is reflected in widely varying replication results across FCs, an outcome of the continuous, intensive, and open collaboration with local stakeholders to produce tailored NBS strategies. FCs experienced that working with local communities requires continuous iterations between analyzing potential sites/environments and refining the transformation vision while detailing all planned NBS.

The roadmap has demonstrated its usefulness for cities, experts, and communities engaged in NBS co-design projects. Admittedly, municipalities may find it challenging to engage in co-design processes often being confined by tight timelines and budget constraints. However, the flexible roadmap tool allows cities/experts to adapt co-design activities accordingly, if urban planning frameworks and local mindset recognize the value of NBS integration.

FC portraits: outcomes of the co-creation process and replication strategies

Cascais: reclaiming fragmented lands and connecting citizens with nature

FC Cascais portrait

BASIC INFO

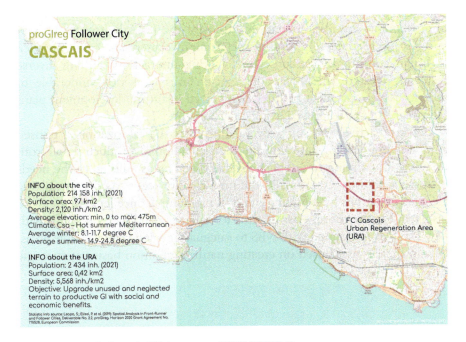

Figure 16.6 FC Cascais URA context (URBASOFIA).

WHY READ THE CASCAIS STORY?

Cascais' journey illustrates how small-scale, cost-effective interventions can spark the revitalization of an entire neighborhood impacted by fragmented or unplanned developments. If you're representing a city or community grappling with challenges such as poor pedestrian accessibility to existing GI assets, difficulty generating intervention ideas for vacant lands, land ownership barriers, or obstacles in implementing urban gardening policies, then Cascais' experience offers valuable insights. It demonstrates how strategically phased interventions can transform a fragmented urban environment into a vibrant, nature-inclusive neighborhood, shielding it from pollutants. By replacing gray infrastructures with new green assets, communities can create a more sustainable and enjoyable living environment.

WHAT IS CASCAIS FIGHTING WITH?

The challenges in the URA include addressing fragmented built areas, tackling illegal occupations and abandoned buildings, and enhancing ecological value through GI. Flood prevention measures and accessibility improvements are also crucial. Additionally, the URA houses immigrant families and requires solutions for social empowerment and cohesion, such as urban community gardens and fostering interactions among diverse residents. Despite misconceptions, social housing is not vulnerable to security issues and must be directly involved in future development plans for a peaceful and improved living environment.

Cascais is actively pursuing several objectives: the valorization of unused land with high ecological value and the protection of productive soil from urbanization, aiming to establish diverse green areas tailored to the needs of the population. Additionally, the municipality seeks to address existing illegal occupations by initiating removal procedures once legal matters are resolved. Furthermore, Cascais aims to replace abandoned buildings with new facilities integrated with NBS to safeguard ecologically valuable areas. Finally, the city is focused on creating natural retention basins at strategic

Figure 16.7 FC Cascais URA conditions (informal gardening) (CASCAIS).

Unlocking potential 259

Figure 16.8 FC Cascais URA conditions (derelict sites with debris) (CASCAIS).

Figure 16.9 FC Cascais URA conditions (fragmented urban tissues) (CASCAIS).

locations along streams, which serve as green meadows during the dry season and provide recreational opportunities for much of the year.

Urbanistic considerations and co-creation process overview

URA CONDITIONS

The Cascais URA spans 0.42 km^2 and features a densely built environment, traversed by a major road forming part of a vital road infrastructure connecting Cascais and Lisbon, effectively dividing the urban area into north and south sections. The Marianas stream, serving as a crucial blue infrastructure, links the natural spaces within the URA. However, these areas are undervalued, facing mounting pressure for urbanization. Despite Cascais' reputation as a high-quality tourist destination along the coast, the designated regeneration area lies on the outskirts of the municipality. Here, an unplanned built environment dominates, with scattered illegal social allotments and dwellings interspersed between the historic rural villages of Tires and Zambujal in the São Domingos de Rana parish. The lack of public green spaces, leisure amenities, and pedestrian connections is notable, although the local open-air market draws crowds on Saturday mornings. Approximately 19% of the protected ecological areas along the Marianas River corridor consist of underutilized land and abandoned farms amid residential areas (Câmara Municipal de Cascais, (2015) PDM Cascais. https://www.cascais.pt/area/plano-diretor-municipal-0). The river valley faces high flooding risks during the rainy season and transforms into a dry riverbed during the summer months, typical of the Mediterranean climate. Real estate pressure has led to the abandonment of former agricultural lands, while spontaneous and illegal vegetable gardens have emerged near social housing settlements, where 18% of the URA's population resides. Nevertheless, there is a growing public awareness and interest in the ecological significance and recreational potential of the Marianas River, signaling a shift toward recognizing and preserving its value.

KEY OUTCOMES AND LEARNINGS FROM THE CO-CREATION PROCESS

In the first phase of preparatory work, the key outcome was obtaining political approval for the vision of transforming the URA. The Cascais team faced challenges in implementing community-oriented activities without support from the municipality, particularly in the political sphere. However,

Figure 16.10 FC Cascais Marianas river (CASCAIS).

the implementation of a community garden pilot project, alongside other proGIreg local events, sparked productive debates and planning processes for green transformation. Partnerships were forged with the Municipal Environment Department's Rivers Division to conduct studies on the Marianas River, potentially leading to the expropriation of private land near the river margins for future interventions. Collaboration with the Municipal Participation Department proved valuable for engaging citizens and collecting input for implementing NBS.

In the second phase of planning the URA transformation through scenario building, stakeholders expressed significant interest in rehabilitating the green river corridor, emphasizing the creation of a new pedestrian trail. Suggestions included the recovery of the stream and the establishment of a blue/green corridor with additional green leisure areas nearby. Stakeholders also advocated for multi-use green spaces with playgrounds, fitness equipment, and community gardens in vacant lots near residential areas, stressing the need for more natural elements and improved pedestrian connectivity. Participants endorsed proposed solutions and anticipated tangible outcomes, with some

262   Nature-Based Solutions for Urban Renewal in Post-Industrial Cities

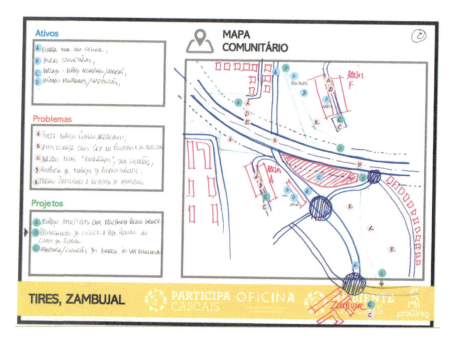

Figure 16.11 Outcomes of FC Cascais co-creation workshops (CASCAIS).

projects potentially funded through participatory budgets. Community maps facilitated a shared understanding of URA assets.

Third stage of co-design focused on transferring the Final Urban Plan proposals and strategic measures to the community and ensuring political commitment. For FC Cascais a primary objective was to raise awareness and present compelling measures for the green transformation of URA. Long-term projects are complex and require policy and regulation measures depending on the revision of the municipal Master Plan. More than often, peripheral neighborhoods like Tires area in Cascais are not in the political agendas of decision-makers, however, through the co-design Urban Plan Cascais team aims to prove feasibility of the interventions, setting a new standard for the pedestrian accessible blue-green corridors, potentially replicable in other areas (Marianas River is crossing multiple neighborhoods). The impact of this corridor is multifaceted: besides the enhanced ecosystemic services, the "frozen" privately owned sites are now accessible and can be the subject of the land take policy for urban agriculture purposes (a local legislative measures very well designed, but with little impact in these type of settings due to no other ancillary leverage mechanism).

FC Cascais Final Urban Plan

The FC Cascais Final Urban Plan emphasizes clustering temporary interventions with critical investments in the 0–5-year period. This includes detailing green corridors and pollinators, along with soft interventions on "blocked" plots. Challenges include ensuring compliance with NBS requirements for the blue-green corridor along Mariana's River. The new improved green corridor systems bring the two communities north and south of the highway together, and the already existing community garden is more accessible and better integrated into the urban tissue. This green spine adds value to apparently invaluable sites. Thus, the artifacts of unplanned developments—sites with no connectivity, become valuable assets for connecting citizens with nature: through landscape and physical activities (urban agriculture activities). The plan also addresses the need for security and protection from the highway and minimal interventions in other areas by replacing concrete with permeable surfaces. Flexibility in design is crucial due to administrative uncertainties, with project fiches accommodating various solutions while providing zoning regulations for parameters related to GI connectivity and multifunctional NBS 3 development (including leisure, sport, and social interaction, in a nature-driven environment).

Figure 16.12 FC Cascais Urban Plan (URBASOFIA).

264 Nature-Based Solutions for Urban Renewal in Post-Industrial Cities

Cultivating change: Cluj-Napoca's approach to green infrastructure development and urban gardening

FC Cluj-Napoca portrait

BASIC INFO

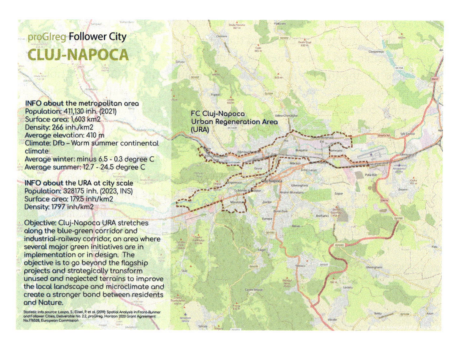

Figure 16.13 FC Cluj-Napoca URA context (URBASOFIA).

WHY READ THE CLUJ-NAPOCA STORY?

Cluj-Napoca's journey in developing strategies for integrating NBS into challenging urban environments, including densely built areas, collective housing districts, industrial and railway axes, and inaccessible blue-green corridors serves as a pertinent example of how relatively small-scale interventions can be woven into a complex transformation process. The urban plan is characterized by its adherence to the urban acupuncture principle, which identifies key areas for intervention, reconnection, and regeneration, such as green plots within collective housing areas and neglected green spaces. Community involvement is a focal point for Cluj-Napoca, with efforts aimed at promoting cohesive communities and encouraging positive environmental behavior. The city emphasizes ecological education, awareness-raising, and sustainable civic engagement to enhance citizens' well-being through nature. A distinctive outcome of the Cluj-Napoca process is the development of an extensive policy proposal for community gardening. This proposal outlines four different typologies of community gardens, each paired with a comprehensive set of zoning regulations tailored to the landscape and morphological conditions of collective housing districts. The document includes design requirements, urbanistic considerations, and proposed business models for the sustainable operation and longevity of the gardens. While the studied area is not unique, as similar developments from the 1960s to 1980s period exist in cities across Romania and the EU, the policy document has the potential to inspire other proponents of green-oriented initiatives in policymaking for scaling up community gardens in dense urban environments.

WHAT IS CLUJ-NAPOCA FIGHTING WITH?

Cluj-Napoca is facing several challenges in its URA. These challenges include enhancing the environmental quality and accessibility to secondary blue-green corridors like the Nădaș River, particularly in post-industrial neighborhoods, to provide residents with access to new green facilities. Protecting the corridor and its surrounding natural areas to ensure biodiversity and mobility is another priority. Given the city's relatively dense urban environment, strategic placement of local interventions in relation to existing GI is crucial. The preferred approach for triggering an "NBS revolution" is urban acupuncture, targeting key areas for intervention, reconnection, and regeneration, such as green plots within collective housing areas and neglected green spaces. In terms of community involvement, Cluj-Napoca is focused on promoting cohesive communities and fostering good environmental behavior. Ecological education, awareness-raising, and sustainable civic engagement are also emphasized to improve citizens' well-being through nature. Additionally, the strategy and urban plan put forward regulatory and policy measures for implementing NBS 3 across various locations. It also includes retrofitting requirements for NBS 5.

Urbanistic considerations and co-creation process overview

URA CONDITIONS

Stretching from east to west, along the river corridor and railway-industrial corridor (double by secondary river network), the chosen URA grapples

Figure 16.14 FC Cluj-Napoca informal vegetable garden (URBASOFIA).

Figure 16.15 FC Cluj-Napoca improvised urban furniture (URBASOFIA).

Unlocking potential 267

Figure 16.16 FC Cluj-Napoca blue-green corridor (URBASOFIA).

Figure 16.17 FC Cluj-Napoca high-traffic street (area for potential green corridor) (URBASOFIA).

with a highly diverse urban environment, comprising fragmented urban tissue from former industrial areas, residential neighborhoods, individual housing, natural corridors, and pertinent infrastructural axes such as the railway line and future mobility corridors. A primary challenge of the co-design process was assessing the urban tissue's capacity to accommodate new interventions. To better understand the study area's characteristics, the URA was divided into seven landscape units.

Understanding the urban structure entails a meticulous analysis of various landscape units within the city, each presenting unique identities and distinct challenges. Beginning from west to east, Area 1 is the Lower Someş Catchment Area, serving as a significant water catchment zone with valuable natural landscapes and ecological diversity, despite its peripheral location. Moving eastward, we encounter Area 2, the Mănăştiur and Plopilor Neighborhood, bustling with urban activity and public amenities, where the sloping land offers potential as a green ecological corridor.

Further along the railway line in the northern part of the city lies Area 3, the Dâmbu Rotund Neighborhood, also known as the West Industrial Area. This area exudes an industrial urban vibe, marked by industrial and commercial activities, with green patches and the Nadăş creek adding to

Figure 16.18 FC Cluj-Napoca aerial image of key intervention area in the URA (ADIZMC).

its landscape diversity. Central to the URA is Area 4, the Iris Neighborhood Residential Area, featuring residential housing with convenient access to the city center. The southern region blends industry, commerce, and services, incorporating abandoned industrial heritage and green-blue corridors.

Adjacent to the residential area is the Iris Neighborhood industrial zone (Area 5), primarily housing industrial activities intersected by the northern bank of the green-blue corridor Someş. Finally, in the southern part of the river lies the Bulgari-Someşeni Area, a complex and densely built zone with diverse functions, divided by the railway into distinct north and south regions.

KEY OUTCOMES AND LEARNINGS FROM THE CO-CREATION PROCESS

Phase 1 yielded significant outcomes for FC Cluj-Napoca, including the creation of a comprehensive assessment of the URA. These activities helped establish the main purpose and role of the URA's GI, connecting it with the city and its natural surroundings. As a result, important new questions emerged, such as identifying key drivers for NBS adaptation, determining specific requirements for community-oriented interventions, assessing the potential impact of each planned NBS, and identifying the necessary resources. The first phase of the co-creation process was critical for defining the URA limit and key intervention areas (with the role of lighthouse projects to establish a new local standard).

In phase 2 of the co-creation process, FC Cluj-Napoca deployed a series of thematic activities to grasp the complexity of the URA priorities in respect to each chosen NBS. Considering NBS 3, residents expressed concerns about potential vandalism and maintenance responsibility, coupled with a low level of understanding of NBS benefits. Although communities welcomed community gardens if properly planned, existing legislation prohibited vegetable growing in public spaces. Transgenerational involvement emerged as a potential strategy, with the elderly passing on knowledge of vegetable growing to younger generations. Additionally, the newly adopted development strategy of Cluj-Napoca could serve as a driver to integrate NBS 3 in future projects. From the point of view of LAN (Local Actors Network) in the first round of discussion in phase 1, NBS 5 started as an uncertain topic for URA transformation but emerged as a priority for buildings retrofitting and achieving climate neutrality. With assistance from the European Green Roof Association, perceptions shifted as stakeholders understood the benefits of green roof solutions. While small-scale initiatives were experimented with, upscaling appeared unlikely. Adapting local policies generated debates, with zoning regulation and future renovation requirements being perceived as the best option forward. For NBS 6, Green Corridors, ensuring connectivity and accessibility of valuable green spaces within the URA emerged as a key aspect. The adaptation of NBS was seen as a step-by-step process, accommodating immediate community needs and favoring modular solutions. Using native plants for minimal maintenance and improving permeability and accessibility of blue-green corridors were highlighted strategies. Local NGO interest in making the lower part of Someş accessible to residents further underscored the importance of community involvement and strategic planning for green corridor interventions.

Phase 3 focused on refining the project proposals resulting from the intensive local co-design process and completing the missing gaps through analysis and citizen-level questionnaires. A notable outcome of Phase 3 is the unplanned delivery of an extended policy proposal elaborated in the local language for the insertion/strategic conversion of green pockets found in collective housing districts into community gardens.

FC Cluj-Napoca Final Urban Plan

The primary objective of FC Cluj-Napoca's Urban Plan is to facilitate the green revitalization and transition of post-industrial neighborhoods and peripheral residential zones. Instead of focusing solely on flagship projects, Cluj-Napoca is prioritizing local efforts to reduce social discrepancies, empowering residents through sustainable green NBS interventions and regenerating local landscapes. Key principles include acknowledging the potential snowball effect for small-scale initiatives, especially in collective housing neighborhoods, and proposing a forward-looking and proactive approach aligned with local urban regulations and plans. This ensures that new developments prioritize inclusive and consistent citizen access to blue-green corridors, making them accessible, sustainable, and imbued with a distinct natural ambiance.

There are three intervention domains:

(1) Converting derelict and underused sites: NBS 3 serves as a flexible greening instrument for converting key sites, setting a new local standard for multifunctional community gardens. These gardens can be

further upscaled at different levels in collective neighborhood areas. This involves transforming underused and informal interventions in green pockets into local neighborhood-level community gardens, following the requirements and recommendations of the policy proposal created specifically for this environment.
(2) Increasing the green spaces network connectivity and mitigating the heat island effect: NBS 6 is a key solution for the peripheral neighborhoods of Cluj-Napoca, linked to underdeveloped blue corridors and mobility infrastructure such as railway embankments, residential streets, and high-traffic areas. The main objective is to establish a continuous green network from the inner city toward the natural surroundings of Cluj-Napoca.
(3) Implementing new requirements for building renovation: The strategy advocates for transitioning from gray to GI. Considering the Renovation Wave, the strategy proposes mainstreaming green roofs for collective housing and public buildings undergoing thermal rehabilitation. This approach aims to achieve energy savings, improved conformity, and biodiversity enhancement.

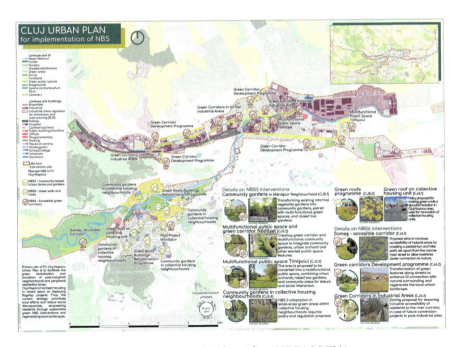

Figure 16.19 FC Cluj-Napoca Final Urban Plan (URBASOFIA).

Piraeus: Gray yards, green yards!

FC Piraeus portrait

BASIC INFO

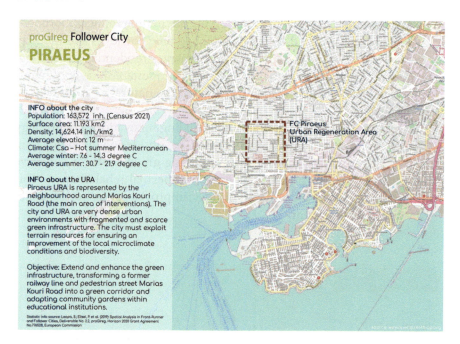

Figure 16.20 FC Piraeus URA context (URBASOFIA).

WHY READ THE PIRAEUS STORY?

Piraeus city presents an extreme case of a densely built and predominantly mineral urban environment, characterized by its hippodamic-specific morphology, which allows minimal space for greenery due to the orthogonal composition of streets. However, Piraeus's involvement in the proGIreg project highlights the significance of investing in greening initiatives, particularly in schoolyard premises and the deployment of green corridors along abandoned infrastructures and derelict public spaces.

Notably, Piraeus's experience underscores the transformative potential of school gardens, which emerged as popular and impactful initiatives. Initially planning to replicate successful strategies and develop a comprehensive strategy, the Piraeus team shifted its focus when approached by schools seeking to address greening needs. Consequently, the municipality opted for small-scale pilots in three schoolyards, involving students in the replanting of derelict areas. This initiative garnered interest from other schools in the URA, prompting the municipality to initiate the development of a local policy proposal aimed at greening schoolyards across Piraeus.

WHAT IS PIRAEUS FIGHTING WITH?

Piraeus faces a multitude of challenges within its URA, including the consolidation of existing GI and the reconstruction of missing linkages, such as Marias Kouri Street. Biodiversity conservation and understanding the connections and impacts of activities on GI and the environment are also key challenges. Additionally, Piraeus seeks to alleviate urban density pressure by integrating natural elements into the urban fabric, improve urban environmental quality, and enhance the quality of public spaces in terms of aesthetics, comfort, and hygiene. The city aims to enhance the multifunctionality of green areas, improve accessibility to public spaces,

Figure 16.21 FC Piraeus MKR corridor (URBASOFIA).

Figure 16.22 FC Piraeus MKR corridor (URBASOFIA).

Figure 16.23 FC Piraeus schoolyards (URBASOFIA).

Unlocking potential 275

Figure 16.24 FC Piraeus schoolyards (URBASOFIA).

and enhance security. Involving local communities is essential, with a focus on raising awareness of the benefits of NBS, ensuring citizen participation in decision-making processes, and engaging young people in planning and design to foster a sense of citizenship and replicate good practices. Improved coordination within the municipality and between departments is also a priority.

Urbanistic considerations and co-creation process overview

URBAN REGENERATION AREA

Follower City Piraeus, the third largest city and municipality in Greece, is situated 12 km southwest of the capital Athens, boasting a population of 163,688 and spanning approximately 11 km^2 in surface area (Hellenic Statistical Authority, 2011a). Renowned for its prominence, the port of Piraeus stands as Greece's most significant and one of the eastern Mediterranean region's crucial ports (Municipality of Piraeus, 2018). However, Piraeus' urbanization trajectory over recent decades has led to the creation of one of

Europe's most densely built cities, contributing to environmental degradation. Presently, the city faces several spatial and environmental challenges, including a dearth of open green spaces and parks, high building density, insufficient infrastructure, and air pollution stemming from ship emissions. The identified URA in Piraeus lies within District City E', situated on the mainland near the ferry port to the south. This district primarily comprises residential areas with small local neighborhood commercial zones, alongside post-industrial areas awaiting regeneration initiatives. Key areas for urban regeneration planning efforts include Marias Kiouri Street, featuring the former tram line and the discontinued light rail track (Piraeus–Perama) since 1977.

KEY OUTCOMES AND LEARNINGS FROM THE CO-CREATION PROCESS

Phase 1 of the co-creation activities uncovered the significant potential of schoolyards for contributing to the greening of the URA, alongside the strategic transformation of Marie Kouri road into a green

Figure 16.25 FC Piraeus schoolyard planting event (URBASOFIA).

corridor. Participants identified key priorities, including the reorganization of parking spots, improvement of accessibility and cycle paths, and the enhancement of school yards through tree planting, orchards, and energy-efficient solutions. Moreover, there was a notable lack of awareness about existing environmental programs in schools, with proGIreg recognized as a significant initiative guiding educational change. The schoolyard transformation theme continued in Phase 2 for scenario building, with the primary outcome -> the collaborative design efforts and school pilot projects, focusing on NBS 3 and NBS 8, was the heightened awareness and sense of ownership among school staff and students regarding the small-scale interventions. This resulted in the initiative spreading to other schools not directly participating in proGIreg activities, highlighting the value of integrating experiential learning programs into school curriculums. For the green corridors in the URA, the Piraeus proGIreg team worked with students to elaborate a series of intervention ideas. In Phase 3, the Piraeus team designed project fiches for school gardens and green corridors, which were then presented in the final event. One of the most notable achievements of the proGIreg project is the increased awareness of the transformative power held by the courtyards of public institutions, particularly in the context of a densely built urban environment.

FC Piraeus Final Urban Plan

In crafting the Final Urban Plan, FC Piraeus faced the primary challenge of integrating a more strategic approach for the medium and long term. Initiatives proposed for Marie Kourie Road, involving the implementation of NBS 6 and NBS 8, are deemed feasible within the 0–5-year timeframe. However, the inclusion of NBS 3 in local schools in Piraeus depends on establishing a municipal policy where the municipality collaborates with schools, offering small grants, funding, or assistance to develop these gardens. The plan emphasizes detailed requirements and design guidelines for the proposed interventions, which were developed through prior co-design activities involving local students. These solutions have been presented to local stakeholders, with further refinement occurring. Strategic directions and guidelines have been provided for replicating specific interventions in other locations. The Final Urban Plan clusters interventions within the 0–5-year timeframe, while the 5–15-year period aims to extend interventions to additional sites within the URA and the city.

Figure 16.26 FC Piraeus Final Urban Plan (URBASOFIA).

Zenica: Activating places and spaces through NBS strategies

FC Zenica portrait

BASIC INFO

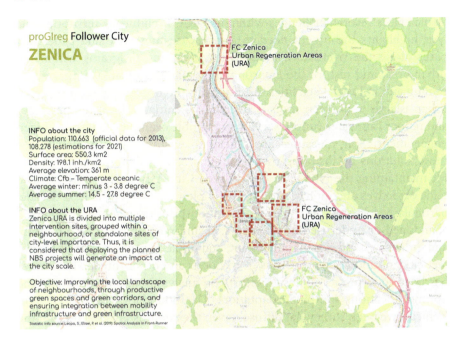

Figure 16.27 FC Zenica URA context (URBASOFIA).

WHY READ THE ZENICA STORY?

Zenica's story reveals the significant value of effective collaboration between municipal departments and between the project team and local actors. The scope of NBS adaptation in the Zenica case has evolved since the project's inception, because of these discussions. Initially, the primary focus was on adapting NBS along the Bosna River. However, consultation sessions led to a different approach, akin to the Cluj-Napoca case, which targeted various key sites, many of which were subject to flagship project designs aimed at dramatically altering the local landscape and establishing a new identity for the city. Zenica is viewed as a close follower of the Zagreb LL, drawing inspiration from the therapeutic garden (NBS 3) piloted by the Zagreb team and developing their own project plan for a multifunctional public space

280 Nature-Based Solutions for Urban Renewal in Post-Industrial Cities

that includes a therapeutic garden directly connected with disadvantaged youth from the vicinity of the site.

WHAT IS ZENICA FIGHTING WITH?

Zenica faces several challenges within its URA. Preliminary territorial analysis highlights the urgent need to connect existing cycle and pedestrian paths along the riverbanks, intensify tree planting programs within the urban area, and restore the environmental qualities of areas impacted by mining and industrial activities. Additionally, the implementation of green urban islands and green roofs is essential to mitigate local temperatures and reduce rainwater runoff, serving as tangible examples to inspire citizen engagement. Soft measures are crucial to ensure the sustainability of these solutions by fostering ownership and responsibility among citizens for maintaining the NBS. The city's Green City Action Plan outlines strategic objectives and interventions to be implemented over the next decade, particularly focusing on urban regeneration priorities such as revitalizing abandoned or underused urban spaces, ensuring compliance with housing and greenery standards per inhabitant, and implementing measures to protect citizens and their properties from floods and other threats. FC Zenica has been focusing its co-design efforts on analyzing and determining the optimal locations for NBS adaptation. The boundaries of the URA in Zenica have been somewhat diffuse, given the circumstances. Through the co-design process, the FC Zenica team has identified ten projects in different parts of the city, closely connected to existing GI such as the river corridor, required pedestrian connections, and underutilized green spaces.

Urbanistic considerations and co-creation process overview

URA

The City of Zenica, located centrally in Bosnia and Herzegovina, sits approximately 70 kilometers northwest of the capital, Sarajevo. It is the fourth largest city in the country and serves as the administrative center of Zenica-Doboj Canton, accommodating around 30% of the total 364,433 inhabitants of the canton. With an area of approximately 550 km², Zenica boasts a notable amount of green spaces within its urban landscape, comprising 58% of the total urban area. However, these green spaces lack functionality due to poor organization, presenting an opportunity for the

Unlocking potential 281

Figure 16.28 FC Zenica key intervention area for a therapeutic garden (ZEDA).

Figure 16.29 FC Zenica key intervention area for remediating former landfill and coal mine area (ZEDA).

Figure 16.30 FC Zenica Babina Rijeka park (ZEDA).

Figure 16.31 FC Zenica key intervention area for NBS 3 (ZEDA).

development of quality GI, especially given the city's location in a river valley. The predominantly flat terrain facilitates connectivity through various modes of transportation, such as a bike-sharing system, with potential for expansion and interconnection of existing cycle and pedestrian paths. Engaging local communities in implementing NBS, like community urban gardening, offers a promising avenue for involving diverse stakeholders and addressing issues such as therapeutic gardens. While sporadic initiatives of a similar nature are already underway in neighborhoods, there is a need for more systematic organization. Pilot actions aimed at raising awareness and generating knowledge about the benefits of nature-based solutions will be implemented, guided by the Final Urban Plan developed in the proGIreg

Unlocking potential **283**

project. This plan will help steer the green transition process, leveraging the expertise offered by the project and identifying NBS activities that can be replicated and implemented in various locations within the urban area of Zenica.

KEY OUTCOMES AND LEARNINGS FROM THE CO-CREATION PROCESS

The preparatory work identified several key issues at the local level in Zenica, including a lack of soft mobility connections between riverbanks, fragmented cycle paths, air pollution from industrial activities, a disjointed distribution of green areas, river overflowing, and citizen engagement. To address these issues efficiently, the city opted for dispersed actions across six different key transformation areas rather than focusing solely on a specific location. It emphasized the importance of integrating proGIreg plans into existing planning documents, aiming for synergy between proGIreg objectives and the Green Action Plan. During the URA transformation planning phase, stakeholders validated the vision and consolidated NBS and actions for each key transformation area. Commitment from relevant stakeholders, including NGOs and residents, was secured for planning and future design activities. Regarding specific outcomes for NBS interventions, the involvement of the University of Mostar was secured to support the planning of regenerating the old landfill area, characterized by coal deposits. For NBS 3, abandoned or underused buildings and green spaces were identified as opportunities for integrating urban gardening practices, addressing local issues of reuse and greening. Urban gardens were also planned to serve educational and therapeutic functions, including the recovery of a former kindergarten and integration into the City Park renovation project. NBS 5 outcomes focused on integrating green roofs and walls to support greening interventions alongside urban gardens, with potential educational functions. Additionally, there was consideration for integrating green roofs on private properties, such as garages, involving private individuals in their maintenance. For NBS 6, scenarios were developed to reconnect existing sections of cycle and pedestrian paths along the riverbanks.

FC Zenica Final Urban Plan

The main goal of FC Zenica's Final Urban Plan is to act as an urban revitalization blueprint, embracing bold and complex projects (such as the regeneration of the Pusara landfill, the regeneration of major urban park

Babina Rijeka, or the implementation of a therapeutic garden in one of the main squares of the city). Its core mission is to offer tailor-made design solutions for each site, finely attuned to the local context, with the aim of enriching citizens' daily experiences, enhancing the local microclimate, and bolstering ecosystem services. The FC Zenica team has orchestrated collaborative efforts among municipal departments overseeing existing projects and investments, linking them to the development and co-design of NBS requirements and concepts. The urban plan's scope is to seam-

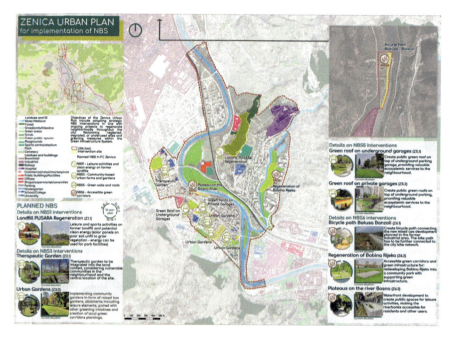

Figure 16.32 FC Zenica Final Urban Plan (URBASOFIA).

lessly infuse NBS into project blueprints, effectively transforming them into invaluable resources for local communities. The Zenica Urban Plan's overarching objectives encompass adapting strategic existing projects to revitalize neighborhoods across the city, reclaiming overlooked, degraded, or underutilized sites, and implementing greening measures within the existing GI network.

NBS policymaking at the forefront 17

NBS for change and resilience

Karin Zaunberger

Nature-based Solutions in policies

Contextualization of the concept

The term 'nature-based solutions' (NBS) was first used in the early 2000s, but has since been widely adopted worldwide and included in relevant policy frameworks seeking to "promote synergies between nature, society and the economy" (Somarakis et al., 2019). NBS have evolved and continue to do so on different levels and scales, from international, regional, national to subnational, to local, and city levels. The concept of NBS is one of several concepts that promote the maintenance, enhancement, and restoration of biodiversity and ecosystems as a means to address multiple concerns simultaneously (Kabisch et al., 2016). All share a common rationale. The idea is to work with nature rather than against it recognizing that healthy ecosystems deliver crucial ecosystem services on which societies and economies depend, involving NBS in rural and urban areas, and in both marine and terrestrial ecosystems. The term NBS has been widely adopted and included in relevant policy frameworks. Ecosystem-based approaches, including green infrastructure (GI), are encompassed within the NBS umbrella concept (Kabisch at al., 2016). These approaches aim to manage land, water, sea, and living resources in a way that promotes conservation and sustainable use in a holistic and equitable way. The NBS

DOI: 10.4324/9781003474869-17

This chapter has been made available under a CC-BY license.

concept is based on a scientific understanding of the interconnectedness of nature and people and prizes biodiversity and functioning ecosystems and their services (supporting, regulating, provisioning, and cultural) within the landscape/seascape (Pörtner et al., 2021).

The concept of working with nature is anchored in many sectors as well as in traditions of indigenous peoples and local communities. This means that there are many different entry points for the implementation of NBS. There has been positive cross-fertilization between advances in research and the design and implementation of policies in different fields involving different actors, including in the examples below.

In the frame of climate change adaptation, ecosystem-based approaches for adaptation have been gaining momentum for over a decade. The Durban Community Ecosystem Based Adaptation (CEBA) initiative (Durban Community Ecosystem-based Initiative, 2017) was among the first ten actions recognized for the Momentum for Change initiative started at the Seventeenth Conference of the Parties to the United Nations Framework Convention on Climate Change (UNFCCC COP17) in Durban in 2011 (Momentum for Change Initiative, 2017). CEBA focuses on the link between communities and the ecosystems that underwrite the welfare and livelihood of these communities.

In the aftermath of the super storm Sandy, which had devastating impact in 2012, the Obama Administration released a memorandum directing federal agencies to factor the value of ecosystem services into federal planning and decision-making (Incorporating Natural Infrastructure and Ecosystem Services in Federal Decision Making, 2015). Ten years later in 2022 the White House issued a roadmap for the accelerated implementation of NBS (Opportunities to Accelerate Nature-based Solutions: A Roadmap for Climate Progress, Thriving Nature, Equity and Prosperity, 2022).

Parties to the Convention of Biological Diversity (CBD) at CBD COP12 in 2014 adopted a decision on biodiversity and climate change and disaster risk reduction (Convention on Biological Diversity Decision 12/20, 2014). The Sendai Framework on Disaster Risk Reduction 2015–2030 adopted in 2015 by World Conference on Disaster Risk Reduction held in Sendai, Japan, and later endorsed by the UN General Assembly, recognized the important role of ecosystems for disaster risk reduction (Sendai Framework for Disaster Risk Reduction, 2015). At CBD COP14 in 2018, Parties adopted voluntary guidelines for the design and implementation of ecosystem-based approaches including environmental and social principles and safeguards (Convention on Biological Diversity Decision 14/5, 2018).

The UN Climate Action Summit in 2019 featured a dedicated work stream on NBS co-led by China and New Zealand (UNEP Authors, 2019).

Since then the number of countries including NBS in their Nationally Determined Contributions (NDCs) has increased (World Wildlife Fund, 2021). However, through the work stream the attention on NBS increased significantly, and unfortunately at the same time the abuse of the concept, for example, labelling monoculture plantation for carbon offsetting with negative impacts on people and biodiversity was wrongly labelled as NBS. This created controversy in many fora where some voices claimed that NBS are vague concept drawing the attention to the lack of a multilaterally agreed definition of NBS.

In 2020, the International Union for Nature Conservation (IUCN) launched a Global Standard for Nature-Based Solutions to help users design, implement, and verify NBS actions (IUCN, 2020).

Parties at UNFCCC COP26 in 2021 recognized the crucial role of nature for climate change mitigation and adaptation with many nature-related events and pledges for NBS. The mitigation section of the Glasgow Climate Pact (United Nations Framework Convention on Climate Change Authors, 2021) 'emphasizes the importance of protecting, conserving and restoring nature and ecosystems to achieve the Paris Agreement temperature goal'. This explicit connection between the climate and nature agendas is more pronounced than in the Paris Agreement from UNFCCC COP21. However, the term NBS was not retained in the final decision text in spite of the support of several Parties.

In 2022 at the United Nations Environment Assembly UNEA 5.2, countries adopted a Resolution on Nature-Based Solutions for supporting Sustainable Development, UNEP/EA.5/Res.5, which provided a multilaterally agreed definition and framing of the concept. It defines NBS as

> actions to protect, conserve, restore, sustainably use and manage natural or modified terrestrial, freshwater, coastal and marine ecosystems, which address social, economic and environmental challenges effectively and adaptively, while simultaneously providing human well-being, ecosystem services and resilience and biodiversity benefits' and states that NbS 'respect social and environmental safeguards.

The resolution also makes it clear that NBS are not a panacea and are additional to rapid reduction of emissions of greenhouse gases. Further countries have requested the Executive Director of the United Nations Environment Programme to carry out an intergovernmental consultation to support the implementation of NBS as defined in the resolution (United Nations Environmental Assembly 5.2 Authors, 2022).

In the same year at UNFCCC COP27, Parties adopted the Sharm El Sheik Implementation Plan (the SHIP) (United Nations Framework Convention on Climate Change Authors, 2022) which includes a reference to NBS in line with the UNEA Resolution UNEP/EA.5/Res.5 on Nature-based Solutions for Sustainable Development. At the same occasion the ENACT Partnership for Nature-based Solutions co-led by COP27 President Egypt and Germany was launched. ENACT – Enhancing Nature-Based Solutions for Climate Transformation – recognizes that this decade represents a critical window for tackling interdependent biodiversity, land degradation, and climate crises and that when implemented properly, NBS can enhance the resilience of ecosystems and the societies that depend on them. NBS can support adaptation to climate hazards such as sea-level rise and more frequent and intense flooding, droughts, heatwaves, and wildfires – while delivering significant biodiversity benefits in a manner that safeguards and promotes the rights and interests of vulnerable and historically marginalized communities (United Nations Framework Convention on Climate Change COP27 Initiative, 2022).

At CBD COP15 under the Chinese Presidency hosted in Montreal in 2022, Parties adopted the Kunming Montreal Global Biodiversity Framework (Convention on Biological Diversity Authors, 2022) called historic by many. The framework reflects never-before-seen recognition from countries at all income levels that loss of biodiversity must be stopped through high-ambition changes to society's relationship with nature and the way our global economy operates. The outcome includes global targets and timetables that – if swiftly, effectively, and efficiently implemented – can see nature pulled back from the brink by 2030, and help keep 1.5°C alive. Two targets, targets 8 and 11, are directly referring to NBS.

Considering the multilaterally agreed definition for NBS as

> actions to protect, conserve, restore, sustainably use and manage natural or modified terrestrial, freshwater, coastal and marine ecosystems, which address social, economic and environmental challenges effectively and adaptively, while simultaneously providing human well-being, ecosystem services and resilience and biodiversity benefits.
>
> (United Nations Environmental Assembly 5.2 Authors, 2022)

implementing NBS means making a significant contribution to achieving the targets agreed in Montreal while at the same time providing decent jobs for many as shown in the recent joint report "Decent Work in Nature-based Solutions" of the International Labour Organization (ILO), the IUCN and the United Nations Environment Programme (UNEP). The report states that 'NbS can generate millions of new jobs, but 'just transition' policies are needed' (International Labour Organisation, United Nations Environment Programme, International Union for Nature Conversation, 2022).

Europe has shown a clear commitment in policy and research to NBS to societal challenges. NBS are an important component of vital strategies for the future health of Europe's people and environment from the Biodiversity Strategy 2030 to the Green Deal (McQuaid et al., 2021). In the European Union (EU), NBS have evolved in different policy areas and sectors sometimes using different terms. The Biodiversity Strategy 2011–2020 (European Commission Authors, 2011) included a dedicated target on GI, and the related GI strategy (European Commission Authors, 2013) promoted GI as a strategically planned network of natural and semi-natural areas with other environmental features designed and managed to deliver a wide range of ecosystem services such as water purification, air quality, space for recreation, and climate mitigation and adaptation. This network of green (land) and blue (water) spaces can improve environmental conditions and therefore citizens' health and quality of life. It also supports a green economy, creates job opportunities, and enhances biodiversity. In the water sector, natural water retention measures (NWRM) have been promoted for water management and flood protection (European Commission Authors, 2015). The European Green Deal (European Commission Authors, 2019) and its related initiatives including, e.g., the EU's Biodiversity Strategy for 2030 (European Commission Authors, 2020) and the EU's Adaptation Strategy (European Commission Authors, 2021) refer to NBS as crosscutting solutions. The EU Research Programme Horizon 2020 (2014–2020) has made significant investments to improve and demonstrate NBS (European Commission Authors, 2023) and the work continues under Horizon Europe (2021–2027). The European Biodiversity Partnership BIODIVERSA+ (Biodiversa+, 2021) sets out to contribute to knowledge for deploying NBS.

NBS can be most effective when planned for longevity and not narrowly focused on rapid carbon sequestration (Pörnter et al., 2021). However, there are limitations to NBS, and their implementation will only be fully effective when they are accompanied by ambitious reductions in emissions. NBS are not a substitute for the rapid phase-out of fossil fuels, and their implementation must not delay urgent actions to decarbonize our economies (Nature-based Solutions Initiative, 2021).

To address the damaging consequences of climate change and biodiversity loss, we need to step up and scale up the implementation of technological solutions and NBS and societal solutions. This requires interdisciplinary teams and knowledge. We need to continue and scale up fruitful collaboration and accelerate and step up the implementation of NBS from small scale to large scale while at the same time further advance our knowledge and work together so that NBS are addressed in a coherent manner in climate and biodiversity and other policy fora. An increasing number of individual countries' climate plans now include NBS. Over 90% of updated government's climate pledges mention nature (World Wildlife Fund, 2021).

Many earlier international reports, action agendas, and policy intentions feature the virtues of NBS. A few examples are listed below.

The UN World Water Development Report on Nature-based Solutions published in 2018 showed that NBS can play an important role in improving the supply and quality of water and reducing the impact of natural disasters (United Nations Water Conference Authors, 2018). One of the three objectives of the Sharm El Sheik to Kunming Action Agenda launched in 2018 at CBD COP14 is to inspire and help implement NBS to meet key global challenges (Convention on Biological Diversity Authors, 2018). In the NBS Manifesto developed for the UN Climate Action Summit 2019, it is stated that: 'NbS are an essential component of the overall global effort to achieve the goals of the Paris Agreement on Climate Change. They are a vital complement to decarbonization, reducing climate change risks and establishing climate resilient societies. They value harmony between people and nature, as well as ecological development and represent a holistic, people-centred response to climate change. They are effective, long-term, cost-efficient and globally scalable …' (United Nations Environment Programme Authors, 2019). The Brazil EU sector dialogue on NBS (2015–2019) concluded that NBS are not only economically smart investment choices, often cheaper than technological solutions, but they can also enhance our quality of life and provide opportunity to shift to a new economy and a new lifestyle more connected to nature (European Commission Authors, 2019). The Leaders' Pledge for Nature (LPN) (Leaders Authors, 2020) was launched in the wake of the United Nation Summit on Biodiversity in 2020. At the moment of writing this chapter the LPN was endorsed by 96 Heads of State or Government and supported by an increasing number of organizations, business leaders, and civil society. The LPN provides a holistic roadmap for action on international level. Leaders commit *inter alia* to move towards a resource-efficient, circular economy, promote behavioural changes, and a significant scale-up in NBS and ecosystem-based approaches on land and at sea. The World Economic Forum (WEF) refers to NBS as an umbrella concept that covers a whole range of ecosystem-related approaches, all of which address societal challenges. These approaches can be placed into five main categories: ecosystem restoration approaches; issue-specific ecosystem-related approaches; infrastructure-related approaches; ecosystem-based management approaches; and ecosystem protection approaches (World Economic Forum Authors,

2021). The UNDRR report "Words into Action Nature-based Solutions for Disaster Risk Reduction" provides a detailed description of the evolution of both concepts, NBS and ecosystem-based approaches and many practical examples (United Nations Office for Disaster Risk Reduction Authors, 2021). While in many fora (CBD, UNFCCC, G20, etc.) fierce debates were held on the relationship between the two terms, the disaster risk reduction community has been using both terms interchangeably.

Transformative change through implementing nature-based solutions

As shown in the preceding section, a lot is happening notably in the area of NBS. However, for now at least we are still in a vicious circle where emission reduction ambition is still too weak to reach the Paris Agreement objectives and increasing biodiversity loss and ecosystem depletion weakens their climate capacity. We can change this by aligning and strengthening climate and biodiversity ambition, we can enter a virtuous circle where strong emission reduction helps to reduce climate impact on ecosystems, which in turn deliver the essential services in which societies and economies depend. At the same time we need to stop excessive human pressure on our ecosystems and biodiversity so that they and we can better cope with the impacts of climate change. The targets 1 to 10 of the Kunming Biodiversity Framework are addressing these drivers of biodiversity loss (Convention on Biological Diversity Authors, 2022).

Limiting global warming to ensure a habitable climate and protecting biodiversity are mutually supporting goals, and their achievement is essential for sustainably and equitably providing benefits to people. Treating climate, biodiversity, and human society as coupled systems is key to successful outcomes from policy interventions. Climate change and biodiversity loss are interdependent and mutually reinforcing. This means that satisfactorily resolving either issue requires consideration of the other (Pörnter et al., 2021).

Both challenges have the same causes. Many of the direct (e.g. changes in land and sea use) and most of the indirect drivers (e.g. unsustainable consumption and production of food, feed, materials, and energy) are the same. Addressing these common drivers must be an essential part of efforts to address both challenges. Land-use change may result in increased greenhouse gas emissions, reductions in sequestration potential, biodiversity loss, and decreased resilience of ecosystems compromising their adaptation capacities and their capacity to deliver essential ecosystem services. Addressing behavioural change and consumption patterns such as excessive consumption of meat would reduce pressures on both climate change and biodiversity (Convention on Biological Diversity Authors, 2019).

The Urban Greening Platform (European Commission Authors, 2022) supports towns in restoring nature and biodiversity including through the implementation of urban NBS. The platform aims to provide guidance and knowledge to support towns and cities in enhancing and restoring their urban nature and biodiversity, along with links to other relevant European Commission initiatives and policies. It features Urban Greening Plan Guidance and Toolkit developed in collaboration with EuroCities, a network of more than 200 cities in 38 countries, representing 130 million people, working together to ensure a good quality of life for all, and ICLEI, Local Governments for Sustainability, a global network of more than 2,500 local and regional governments committed to sustainable urban development which is active in 125+ countries. ICLEI Europe supports local governments in implementing the European Green Deal, the overarching EU strategy for climate neutrality, to build more resilient and equitable communities.

Policies that simultaneously address synergies between mitigating biodiversity loss and climate change, while considering their societal impacts, offer the opportunity to deliver multiple benefits and help meet development aspirations for all. The explicit consideration of the interactions between biodiversity, climate, and society in policy decisions provides opportunities to maximize co-benefits and to minimize trade-offs and co-detrimental (mutually harmful) effects for people and nature. The climate–biodiversity–social system is a nexus most appropriately dealt with from a social ecological perspective. The creation of GI in cities is increasingly being used for climate change adaptation and restoration of biodiversity with climate mitigation co-benefits (Pörnter et al., 2021).

Under the effects of biodiversity loss and climate change, crucial (hard to reverse or irreversible) thresholds (tipping points) can be exceeded with dire consequences for people and nature, but positive social tipping points can help attain desirable biodiversity-climate interactions. Surpassing thresholds can lead to changes in ecosystem function. For example, climate change can cause biophysical limits of corals to be exceeded or sea-ice ecosystems to disappear, leading to regime changes and algal-dominated communities with markedly different function (Pörtner et al., 2021). Coral reefs buffer coasts against storm surges and waves. They absorb over 95% of the waves' energy. Crumbled coral reefs no longer protect shores and coastlines where many cities are located will erode and become inhabitable.

Business plays an important role to accelerate the implementation of NBS both directly through actively using NBS and indirectly through unlocking funding and investments in the implementation of NBS. The We Mean Business Coalition has identified five guiding principles for corporate climate leadership:

Principle 1: reduce emissions in your own business as quickly as possible;
Principle 2: cut your land-based emissions in your value chain;
Principle 3: invest in nature-based solutions beyond your value chain;
Principle 4: ensure responsible policy engagement on climate and nature;
Principle 5: report and communicate transparently.

These principles are a framework for corporate leaders to cut emissions in their value chains and go further in protecting and restoring nature beyond their value chain (We mean Business Coalition, 2022).

The EU-funded project Network Nature facilitates the Connecting Nature Enterprise Platform (CNEP) launched in 2020 as a response to research findings that identified thousands of nature-based enterprises (NBEs) working for and with nature. These enterprises often operate in isolation with little recognition, networking or support. CNEP offers NBEs a platform where they can connect with their peers, learn about good practices and market trends. The platform connects NBEs with potential buyers highlighting their expertise. Today, there are ten dynamic communities of practice led by industry ambassadors who organize regular webinars and activities, and Continuing Professional Development (CPD) courses to keep everyone connected and up to date. At CNEP, we continue to build and expand our resource library where we share knowledge and insights on emerging technologies and opportunities in the nature-positive economy (Network Nature Authors, 2020).

The most effective tool to reversing warming and biodiversity loss is to turn to the people who are most impacted, with the greatest need, yet are the least heard and then listen, support, and empower. The climate and biodiversity crisis will not be addressed unless the bulk of humanity is engaged. Turning the tide of biodiversity loss and climate change will only be possible if our entire society embarks on a deep, rapid, and transformative change from trade and production to our consumption and lifestyles (Hawken, 2021).

The implementation of NBS engages and empowers people. It is labour intensive and requires a multitude of skills both blue and white collar. It requires collaboration across disciplines, and it engages all parts of society and different age groups. Engaging in NBS can bring more nutritious food, clean water, clean air, resilient and profitable agriculture, restored fisheries, and improved public health. Implementation of NBS is not a panacea, but can make a significant contribution to address the planetary emergency we are in. The benefits are multiple: more jobs, more food security, and more resilience.

Participatory budgeting could be a source for implementing NBS. This approach has been pioneered in Brazil and is now used in more than 12,000 locations in the world. Citizen assemblies are another tool to engage citizen in decision-making (Bregman, 2020).

Achieving the scale and scope of transformative change needed to meet the goals of the UNFCCC, and CBD and the Sustainable Development Goals rely on rapid and far-reaching actions of a type never before attempted (Pörnter et al., 2021). We need to address governance challenges by using reflexive approaches, bringing together new networks of society, NBS ambassadors and practitioners, by adapting administrative and legal frames and allocating sufficient budget for implementing and maintaining green space projects in cities. However, it is clear that there will not be a "one size fits all" solution, because this will depend on the local contexts. We need governance approaches that take into account an integrative and interdisciplinary participation of diverse actors. While doing so socio-environmental justice and social cohesion need to be considered making sure that no one is left behind thanks to fair access to green spaces when building more compact urban areas. In growing cities or in cities striving to "infill" urban areas, a common vision needs to be shared and implemented through the right incentives and regulation and should address competing land uses between more houses or buildings and preservation of nature in cities (Kabisch et al., 2016).

Taking these needs into account NBS can serve as climate change mitigation and adaptation tools providing multiple benefits for human health and for societal well-being.

The COVID-19 pandemic has had a deep impact on our lives, but it also has shown that fast and deep change is possible. The pandemic has drawn the attention to the fact that the same human activities that drive climate change and biodiversity loss also drive pandemic risk through their impact on the environment. "Protect and preserve the source of human health: Nature" was the first prescription of the WHO manifesto for a Healthy Green Recovery from COVID-19 (World Health Organization Authors, 2020).

The pandemic has also shed light on the importance of accessible green spaces, urban NBS, for our mental and physical health (EKLIPSE Expert Working Group, 2022). NBS in urban areas provide space for nature and also contribute to addressing the urban heat island effect, carbon sequestration, improved air quality, water management, and flood protection. Thereby,

NBS serve as strong investment options for sustainable urban planning, providing job and business opportunities and contributing to increased resilience and liveability in cities. This relates directly to Target 12 of the Kunming Montreal Framework: 'Significantly increase the area and quality and connectivity of, access to, and benefits from green and blue spaces in urban and densely populated areas sustainably, by mainstreaming the conservation and sustainable use of biodiversity, and ensure biodiversity-inclusive urban planning, enhancing native biodiversity, ecological connectivity and integrity, and improving human health and well-being and connection to nature and contributing to inclusive and sustainable urbanization and the provision of ecosystem functions and services' (Convention on Biological Diversity COP/15/5, 2022).

Considering the above it becomes clear that implementing NBS, while ensuring a just transition leaving no one behind, de facto contributes to the transformative change needed to make sure that our children and grandchildren can thrive. The urban population is growing rapidly. Through massive implementation of urban NBS jobs can be created, livelihoods and health of city dwellers can be improved while at the same time making a significant contribution to achieving climate and biodiversity objectives.

We need to unite behind science and make science-based decisions. Accelerated implementation of NBS can help to achieve real engagement with citizen and achieve transformative change. The labour intensity of implementing NBS which has at times been considered to be a barrier in traditional business plans because of the perceived costs should rather be seen as an asset to provide jobs and improve livelihoods.

Increasing investment in NBS remains crucial for a resilient recovery and sustainable development. We must not lock ourselves into damaging old habits. We need to take urgent, concerted, and collaborative actions to reconnect with nature and build a resilient and sustainable global economy that incorporates nature at its heart, even as or the more so we build back from the COVID-19 crisis and the economic and social shocks it entails. Sadly two years after the pandemic with war and conflict rising, the opportunity until now has largely been missed.

And yet, how we will stimulate the economy and allocate capital will either amplify the planetary emergency or help to address it. We need to stop making near-sighted decisions at the expense of the decisions needed to achieve fast and deep transformative change. Now is the moment to invest in nature, phase out fossil fuels, and move to a circular, net-zero carbon, nature positive, sustainable, and equitable economy. Natural capital investment, including restoration of carbon-rich habitats, climate-friendly agriculture, nature-friendly forestry, and sustainable tourism, is recognized to be among the five most important fiscal recovery policies, which offer high economic multipliers and positive climate impact.

Today, some 56% of the world's population – 4.4 billion inhabitants – live in cities. This trend is expected to continue, with the urban population more than doubling its current size by 2050, at which point nearly seven of ten people will live in cities (Worldbank Authors, 2023). This is both an enormous challenge and a tremendous opportunity. Cities can be the drivers for the fast transformative change which is needed. Increasing investment in urban NBS can make a significant contribution to make transformative change a reality. Most NBS are based on four elements with transformation potential: nature's values, knowledge types, community engagement, and management practices (Palomo et al., 2021).

Final thought: The transition to sustainability is urgent and necessary. The adoption of the Kunming Montreal Global Biodiversity Framework (Convention on Biological Diversity COP 15/5, 2022) has been called historic by many. Now we need to make sure that its implementation will be equally historic. This needs an all-government, all-society, and all-economy engagement. We still have the choice, but the window of opportunity is closing fast. This is the decade for action. Each year, each choice, and each decimal degree matter. We need systemic change and we need it fast, as scientists are warning us that we are getting dangerously close to irreversible cascading tipping points in the earth system potentially leading to a "hothouse earth" unlivable for human beings (Lenton et al., 2019; Brovkin et al., 2021; McKay et al., 2022). Therefore, we need to act boldly now and invest in and implement technological and nature-based and societal solutions to stem the planetary emergency we are in. We know what we have to do and we have the human and financial capital to do it. The question is will we manage to do it in the time available. Nature is the timekeeper. There will be no bailouts with nature. When the ice is gone, it cannot be replanted.

References

Bregman, R. 2020. Humankind A Hopeful History. https://www.theguardian.com/books/2020/may/12/humankind-a-hopeful-history-by-rutger-bregman-review (last accessed 4 January 2022).

Brovkin, V., at al. 2021. Past Abrupt Changes, Tipping Points and Cascading Impacts in the Earth System. https://www.nature.com/articles/s41561-021-00790-5 (last accessed 4 January 2022).

Convention on Biological Diversity (CBD), Decisions CBD COP 14/5; 12/20. 2018. https://www.cbd.int/doc/decisions/cop-14/cop-14-dec-05-en.pdf

NBS policymaking at the forefront: NBS for change and resilience **297**

Convention on Biological Diversity (CBD) Decision COP15/5. Kunming Montreal Global Biodiversity Framework. 2022. https://www.cbd.int/gbf/ (last accessed 4 April 2023).

Convention on Biological Diversity Authors. 2019. https://www.cbd.int/cop/cop-14/annoucement/nature-action-agenda-egypt-to-china-en.pdf (last accessed 3 January 2022).

Convention on Biological Diversity Authors CBD/SBSTTA/23/3. 2019. https://www.cbd.int/doc/c/326e/cf86/773f944a5e06b75dfc5866bf/sbstta-23-03-en.pdf (last accessed 21 January 2023).

Durban Community Ecosystem-based Adaptation (CEBA) Initiative. 2017. https://www.durban.gov.za/storage/Documents/Environmental%20Planning%20%20Climate%20Protection%20%20Publications/Climate/Durban%20Adaption%20Charter%20-%202017%20Annual%20Report.pdf (last accessed 3 January 2022).

EKLIPSE Expert Working Group. 2022. Reports Green and Blue Spaces and Mental Health. https://eklipse.eu/request-health/ (last accessed on 21 January 2022).

European Commission

EU Biodiversity Strategy to 2020 COM/2011/0244 Final. 2011. https://ec.europa.eu/environment/nature/biodiversity/strategy_2020/index_en.htm (last accessed 3 January 2022).

Green Infrastructure Strategy COM/2013/249 Final. 2013. https://ec.europa.eu/environment/nature/ecosystems/index_en.htm (last accessed 3 January 2022).

Natural Water Retention Measures. 2015. https://nwrm.eu/ (last accessed 3 January 2022).

European Green Deal COM/2019/640 Final. 2019. https://ec.europa.eu/info/strategy/priorities-2019-2024/european-green-deal_en (last accessed 3 January 2022).

EU Biodiversity Strategy for 2030 COM/2020/380 Final. 2020. https://ec.europa.eu/environment/strategy/biodiversity-strategy-2030_en (last accessed 4 January 2022).

EU Adaptation Strategy COM/2021/82 Final. 2021. https://ec.europa.eu/clima/eu-action/adaptation-climate-change/eu-adaptation-strategy_en (last accessed 3 January 2022).

The EU and NbS. 2023. https://research-and-innovation.ec.europa.eu/research-area/environment/nature-based-solutions_en (last accessed 3 April 2023).

BIODIVERSA+ 2021. https://www.biodiversa.org/1972/downloadlinks a-h (last accessed 3 January 2022).

EU Brazil Sector Dialogue on Nature-based Solutions. https://ec.europa.eu/info/publications/eu-brazil-sector-dialogue-nature-based-solutions_en (last accessed 4 January 2022).

Hawken, P. 2021. Regeneration: Ending the Climate Crisis in One Generation. https://regeneration.org/ (last accessed 4 January 2022).

Incorporating Natural Infrastructure and Ecosystem Services in Federal Decision-Making. 2015. https://obamawhitehouse.archives.gov/blog/2015/10/07/incorporating-natural-infrastructure-and-ecosystem-services-federal-decision-making (last accessed 3 January 2022).

International Labour Organisation (ILO), United Nations Environment Programme (UNEP), International Union for the Conservation of Nature. 2022. Decent Work in Nature-based Solutions 2022. https://www.ilo.org/global/topics/employment-intensive-investment/publications/WCMS_863035/lang--en/index.htm (last accessed 21 March 2023).

International Union for Conservation of Nature (IUCN). 2020. IUCN Global Standard for NbS. https://www.iucn.org/theme/nature-based-solutions/resources/iucn-global-standard-nbs (last accessed 3 January 2022).

Kabisch, N., et al. 2016. Nature-based Solutions to Climate Change Mitigation and Adaptation in Urban Areas: Perspectives on Indicators, Knowledge Gaps, Barriers, and Opportunities for Action. *Ecology and Society*, 21(2), 39. https://doi.org/10.5751/ES-08373-210239 (last accessed 21 January 2022).

Kabisch, N., Korn, H., Stadler, J., & Bonn, A. (eds.). 2017. *Nature-based Solutions to Climate Change Adaptation in Urban Areas: Linkages between Science Policy and Practice.* https://doi.org/10.1007/978-3-319-56091-5 (last accessed 21 January 2023).

Leaders' Pledge for Nature. 2020. United to reverse Biodiversity Loss by 2030 for Sustainable Development. https://www.leaderspledgefornature.org/ (last accessed 6 January 2023).

Lenton, T.M., et al. 2019. Climate Tipping Points — Too Risky to Bet Against. https://www.nature.com/articles/d41586-019-03595-0 (last accessed 4 January 2022).

McKay, D.I.A., et al. 2022. Exceeding 1.5°C Global Warming Could Trigger Multiple Climate Tipping Points. *Science*, 377, 1171. https://www.science.org/doi/10.1126/science.abn7950 (last accessed 21 January 2023).

McQuaid, S., Kooijman, E.D., Rhodes, M.-L., & Cannon, S.M. 2021. Innovating with Nature: Factors Influencing the Success of Nature-Based Enterprises. *Sustainability*, 13, 12488. https://doi.org/10.3390/su132212488 (last accessed 4 April 2023).

Nature-based Solutions Initiative. 2021. Nature-based Solutions for Climate Change Key Messages for Decision Makers in 2021 and beyond. https://nbsguidelines.info/ (last accessed 3 January 2022).

Network Nature. 2020. https://networknature.eu/ (last accessed 10 April 2023).

Palomo, I. et al. 2021. Assessing nature-based solutions for transformative change https://doi.org/10.1016/j.oneear.2021.04.013 (last accessed 3 January 2025).

Pörnter, H.O., et al. 2021. *IPBES-IPCC Co-sponsored Workshop Report on Biodiversity and Climate Change.* IPBES and IPCC. https://doi.org/10.5281/zenodo.4782538 (last accessed 3 March 2023).

Sendai Framework for Disaster Risk Reduction 2015–2030. https://www.undrr.org/publication/sendai-framework-disaster-risk-reduction-2015-2030 (last accessed 3 January 2022).

Somarakis, G., Stagakis, S., & Chrysoulakis, N. (eds.). 2019. *ThinkNature Nature-Based Solutions Handbook*. Book on a Tree Ltd. https://doi.org/10.26225/jerv-w202.

United Nations Environment Assembly UNEA 5.2. Resolution on Nature-based Solutions for Sustainable Development (UNEP/EA.5/Res.5). 2022. https://wedocs.unep.org/bitstream/handle/20.500.11822/39752/K2200677%20-%20UNEP-EA.5-Res.5%20-%20Advance.pdf (last accessed 4 April 2023).

United Nations Environment Programme Authors. 2019. https://www.unep.org/nature-based-solutions-climate (last accessed 3 April 2023).

United Nations Framework Convention on Climate Change Momentum for Change. 2017. https://sdg.iisd.org/news/unfccc-launches-momentum-for-change-initiative/ (last accessed 4 January 2022).

United Nations Framework Convention on Climate Change UNFCCC Authors Glasgow Climate Pact UNFCCC COP26. 2021. https://unfccc.int/process-and-meetings/the-paris-agreement/the-glasgow-climate-pact-key-outcomes-from-cop26 (last accessed 3 January 2022).

United Nations Framework Convention on Climate Change UNFCCC authors Sharm El Sheik Implementation Plan (SHIP) UNFCCC COP27. 2022. https://unfccc.int/sites/default/files/resource/cop27_auv_2_cover%20decision.pdf (last accessed 21 January 2023).

United Nations Framework Convention on Climate Change UNFCCC COP27. 2022. Initiative Enhancing Nature-based Solutions for Climate Transformation ENACT. BMUV: Egyptian COP27 Presidency, Germany and IUCN announce ENACT Initiative for Nature-based Solutions | Press release and ENACT: Enhancing Nature-based Solutions for an Accelerated Climate Transformation | IUCN.

United Nations Summit on Biodiversity. 2020. https://www.un.org/pga/75/united-nations-summit-on-biodiversity/ (last accessed 4 January 2022).

UN Office for Disaster Risk Reduction. 2021. Words into Action Nature-based Solutions for Disaster Risk Reduction. https://www.undrr.org/words-action-nature-based-solutions-disaster-risk-reduction (last accessed 21 January 2023).

UN Water. 2018. https://www.unwater.org/world-water-development-report-2018-nature-based-solutions-for-water/ (last accessed 3 January 2022).

We mean Business Coalition. 2022. Guiding Principles for Corporate Climate Leadership on the Role of Nature-based Climate Solutions. https://www.wemeanbusinesscoalition.org/blog/guiding-principles-for-corporate-climate-leadership-on-the-role-of-nature-based-solutions/ (last accessed 4 April 2023).

White House. 2022. Report to the National Climate Task Force: Opportunities to Accelerate Nature-based Solutions: A Roadmap for Climate Progress, Thriving

Nature, Equity and Prosperity. https://www.whitehouse.gov/wp-content/uploads/2022/11/Nature-Based-Solutions-Roadmap.pdf (last accessed 21 January 2023).

Worldbank Understanding Poverty Topic Urban Development. 2023. https://www.worldbank.org/en/topic/urbandevelopment/overview#:~:text=Today%2C%20some%2056%25%20of%20the,people%20will%20live%20in%20cities (last accessed 4 April 2023).

World Economic Forum WEF. 2021. What Are Nature-Based Solutions and How Can They Be Harnessed to Tackle Climate Change? https://www.weforum.org/agenda/2021/12/what-are-nature-based-solutions-tackle-climate-crisis/ (last accessed 4 January 2022).

World Health Organization Authors. 2020. WHO Manifesto for a Healthy Recovery from COVID-19. https://www.who.int/news-room/feature-stories/detail/who-manifesto-for-a-healthy-recovery-from-covid-19 (last accessed 21 January 2023).

World Wildlife Fund WWF. 2021. https://wwf.panda.org/wwf_news/?4238891/NDCS-nature (last accessed 3 January 2022).

Business models in NBS 18

Bernd Pölling and Rolf Morgenstern

Introduction: What is this chapter about?

Nature-based solutions (NBS) can exploit their full potential for improved city sustainability and resilience only when considering the economic dimension, preferably being economically viable in the end. In proGIreg, NBS on post-industrial sites are co-designed, co-implemented, and co-managed by local communities and organizations, like NGOs, public authorities, associations, businesses including community-based start-ups, and citizens. This transdisciplinary setting promotes social innovation and circular economy approaches (see earlier chapters). The development capacity of preferably self-sustained business models goes beyond purely economic figures by integrating biodiversity, health, social cohesion, etc. into holistic processes. NBS business models do not necessarily have to be economically viable as long as the overall benefits for society and environment are considered higher than the associated costs. However, the costs have to be covered somehow and by someone. Thus, suitable financing models for NBS are relevant and should be integrated into business model thinking of NBS.

The overarching aim of proGIreg is the integration of eight NBS into (at least partly) self-sustained business models. For reaching this aim, the introduction of business model thinking and business management tools into NBS development is elementary. This requires a mind-set change: traditionally, business thinking is of utmost importance in industry and the wider economic sector only, but not per se in public, community, and private sectors. The transdisciplinary approach of beneficial NBS implementation by community, public, and private sector – ideally together – expands the economic

DOI: 10.4324/9781003474869-18

This chapter has been made available under a CC-BY license.

dimension accordingly. Along with shrinking financial endowments of municipalities and other public entities, it is becoming increasingly relevant to integrate business thinking inherently.

This chapter presents business models as well as their organizational and governance structures implemented in proGIreg's Living Labs, including public-private partnerships, social entrepreneurship, and shift from one-off purchase to recurring revenue stream models. Before focusing on these, barriers and challenges that impede wider NBS implementation are presented.

Barriers for NBS implementations

The path to successfully develop NBS on post-industrial sites in a co-designed, co-implemented, and co-managed manner is peppered with barriers and challenges. In the four project's Front-Runner Cities Dortmund, Turin, and Zagreb in Europe, as well as Ningbo in China, a wide range of barriers occurred; some even stopped further implementation or demanded a shift to an alternative or adjusted type of NBS.

The detection of barriers is based on attending the implementation process from the start. Following a first round of individual interviews, NBS-specific workshops create the second source of information on barriers that occurred during co-design, co-implementation, and co-management of NBS. The workshops took place around a year after the interviews to monitor the barriers' evolution. However, the findings have to be seen in light of varying stages of NBS development: While a continuously increasing number of NBS were implemented, other NBS developments were still in their co-design phase prior to implementation.

Four main types of barriers could be detected in proGIreg NBS developments. These are:

- administrative and institutional barriers;
- technological barriers;
- social and cultural barriers; and
- financial and market barriers.

The empirical work revealed that certain types of barriers dominated in specific phases of the NBS development (see Figure 18.1). Administrative and institutional barriers are mainly challenging the co-design planning phase. Technological barriers also occur during the co-design phase, but increase in severity later on when reaching the implementation phase.

Business models in NBS   303

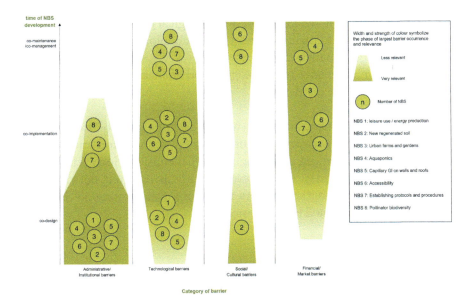

Figure 18.1 Visualization of the occurrence and relevance of barrier types; the numbers in circles indicate the NBS concerned within proGIreg's Living Labs.

Apart from technological barriers, which are often very NBS-specific, financial barriers also emerge progressively with NBS advancement. Only when decision-making bodies (policy, legislation, and administration) become aware of the different types and categories of barriers, will it be possible to support implementation of productive green infrastructure and NBS.

Administrative and institutional and technological barriers prevail in the planning/co-design phase of NBS developments of "new soil", "aquaponics", and "capillary green infrastructure" (green walls and roofs). This dominance of administrative and institutional barriers is no longer true for the implementation and maintenance phase of NBS developments. Suitable and well-defined implementation plans and co-design processes allow administrative and institutional barriers to be overcome, including permissions from municipalities, which are also required. During implementation, technological barriers (which are often NBS-specific) and financial barriers prevail. Apart from the period of NBS development some barriers occur specifically for one or few NBS, while other barriers are identified more often, allowing the conclusion that these barriers tend to be of overarching nature. Barriers that apply to several NBS are:

- bureaucracy/lengthy municipal processes (administrative/institutional barrier);

304 Nature-Based Solutions for Urban Renewal in Post-Industrial Cities

- soil contamination and pollution (technological barrier);
- lack of expertise, knowledge, and skills (technological barrier);
- limited budget (financial barrier);
- long-term maintenance (technological and financial barrier); and
- COVID-19-related restrictions.

Administrative and institutional as well as technological barriers can be overcome by certain measures, including expert consultation and building trans-disciplinary alliances, including municipal entities. However, suitable business ideas and business model developments are required, both for overcoming financial and market barriers as well as for integrating NBS into (at least partly) self-sustained business models.

Business model thinking

Business model concepts to set up and analyse enterprises and organizations have risen since the mid-1990s, although the first appearance dates back to the 1960s (Osterwalder, 2004; Henriksen et al., 2012). Yet, the wider use of the term "business model" is a relatively young phenomenon that has found its first peak during the web-hype at the beginning of the third millennium (Osterwalder, 2004).

Referring to definitions of the terms "business" and "model", Osterwalder concludes 'a representation of how a company buys and sells goods and services and earns money' (Osterwalder, 2004: 14) as a first simple understanding of the term "business model". It aims to support the understanding, description, and prediction of buying and selling goods and services to earn money. A range of different definitions and interpretations exists. Business models explain how companies do businesses (Henriksen et al., 2012); they

- describe 'the rationale of how an organization creates, delivers and captures value' (Osterwalder and Pigneur, 2009: 14);
- stand for the 'design of organizational structures to enact a commercial opportunity' (George and Bock, 2011: 83f.);
- show 'how a firm is able to earn money from providing products and services' (Boons and Lüdeke-Freund, 2013: 9); and
- explain 'how value is created for the customers and how value is captured for the company and its stakeholders' (Henriksen et al., 2012: 31).

Business model concepts have emerged on a system-level dimension as a relatively new addition. They aim to explain holistically how firms operate and do business. Organizational activities play an important role in the various conceptualizations of business models, which seek to explain how value is created and captured. The identification of "who", "what", and "how" is essential for the analysis of business models (Henriksen et al., 2012).

Value proposition, supply chain, customer interface, and financial model are four generic components to be viewed at when analysing business models (Boons and Lüdeke-Freund, 2013). They are suitable not only for getting insights about value creations, relationships, and success factors, but also for comparison with competitors. They consist of interlocking elements that, taken together, create values; for example, customer value propositions and profit (Johnson et al., 1996). Business models allow setting up a supportive overview of how to create and capture value and support knowledge creation and awareness to identify required changes to keep a competitive advantage or for future innovations.

Although the business model thinking stems from the business world, it can be utilized for activities of only partly business-oriented organizations, NGOs, public entities, or others. NBS implementations encompass a wide range of interventions, some of which can create direct monetary values, e.g., by selling products or services or charging fees. Yet as many NBS implementations are led by public entities, like municipalities, many NBS prioritize non-monetary value over monetary value generation. Still, these types of NBS can be integrated into business model thinking and summarized in tools, like the business model canvas (BMC). The European Commission defines NBS as

> as a way to address societal challenges with 'solutions that are inspired and supported by nature, which are cost-effective, simultaneously provide environmental, social and economic benefits and help build resilience. Such solutions bring more, and more diverse, nature and natural features and processes into cities, landscapes and seascapes, through locally adapted, resource-efficient and systemic interventions.
>
> (Faivre et al., 2017)

The economic benefits are equally named and ranked as social and environmental benefits.

Business model tools

The task of capturing business models is an easy one to understand, but at the same time tools have been developed that do not oversimplify this.

306 Nature-Based Solutions for Urban Renewal in Post-Industrial Cities

These tools present business models in a holistic manner. The best-known tool is the BMC developed by Osterwalder and Pigneur (2009). It serves as a widely used strategic management template, summarizing key information on how a business or organization works. Osterwalder, Pigneur, and more than 470 practitioners from 45 countries published "Business Model Generation", in which the BMC is presented in detail. It is a strategic management template to document not only existing business model ideas but also the development and visualization of new ones. BMC is a tool, which provides helpful overviews of companies to emphasize key success factors, to detect barriers, to compare competitors, and to generate business ideas and innovations. The BMC's four main components are customers, offer, infrastructure, and financial viability. Additionally, the BMC template allows working on the desirability, feasibility, and viability of business ideas or business developments. The BMC consists of nine basic building blocks (see Figure 18.2). The positioning of the building blocks allows focusing on:

- the customers and desirability aspects on the right side,
- the value proposition in the central location;
- the infrastructure and feasibility on the left side;
- the financial viability at the bottom; and
- the strategic management template.

Key Partners	Key Activities	Value Propositions	Customer Relationships	Customer Segments
	Key Resources		Channels	
Cost Structure			Revenue Streams	

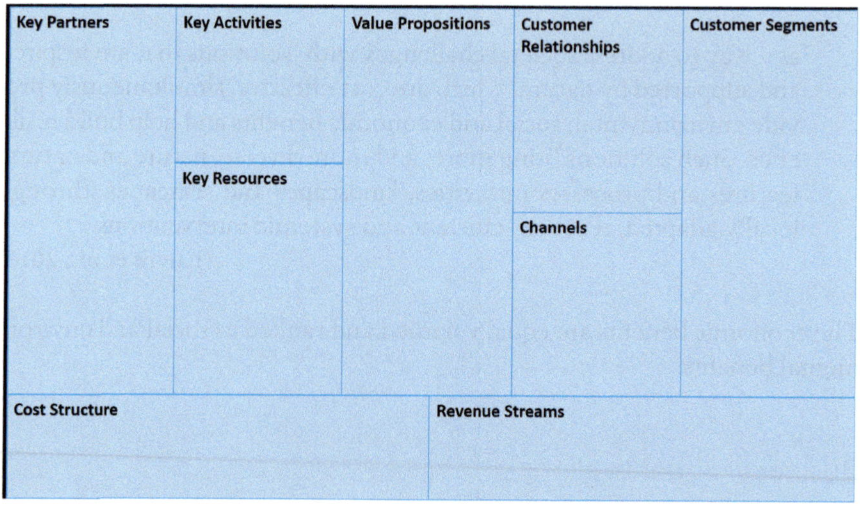

Figure 18.2 Structure of the business model canvas consisting of nine basic building blocks (adapted from Osterwalder and Pigneur, 2009).

The traditional BMC tool, as developed by Osterwalder and Pigneur, has a clear enterprise focus. The sustainability dimension can be integrated into this traditional BMC under value proposition. For example, Ferranti and Jaluzot (2020) use the traditional BMC to increase green infrastructure valuation tools' impact. However, to better consider and represent holistic thinking and sustainability dimensions, several variations and alterations to the traditional BMC have been developed in the 2010s.

One alteration is the so-called triple layered business model canvas that uses one layer for each of the three sustainability dimensions: economic, environmental, and social (Joyce and Paquin, 2016). The economic layer remains the same as the traditional BMC from Osterwalder and Pigneur (2009). The environmental and social layers keep the same structure of nine blocks. This not only results in a horizontal coherence within each of the three layers, but also in a vertical coherence. Left to the centrally positioned social value, the social layer focuses on governance, employees, and local communities summarizing the negative impacts to the bottom left (see Figure 18.3). To the right, the societal culture, scale of outreach, and end user are discussed resulting in the social benefits to the bottom right side of the social layer. The environmental layer follows the same logic: supplies and outsourcing, production, and materials to the left above the environmental impact and end-of-life, distribution, and use phase to the right above environmental benefits. The functional value builds the centrally positioned environmental building block.

Another adaption of the traditional BMC towards emphasizing sustainability is the sustainable business model canvas proposed by Gerlach (2015) (see Figure 18.4). This template aims to incentivize sustainable product and business model design through stronger consideration of all aspects relevant for a holistic business model design (economical, environmental, and sociocultural aspects).

With the emergence of NBS projects and applications in different European cities, business model tools emphasizing specifically on NBS emerged. Two EU projects integrating business model approaches into NBS activities are Connecting Nature (Link) and Naturvation (Link).

The Connecting Nature approach (see Figure 18.5) modifies the traditional BMC (Connecting Nature, 2019). However, it maintains the main concept and structure.

The main changes of Connecting Nature's BMC modification are as follows:

- From customer segments to key beneficiaries: This broadens the consideration of people and entities, who might be customers or direct end

Local Communities	Governance	Social Value	Societal Culture	End-User
	Employees		Scale of Outreach	
Social Impacts			Social Benefits	

Supplies and Out-Sourcing	Production	Functional Value	End-of-Life	Use Phase
	Materials		Distribution	
Environmental Impacts			Environmental Benefits	

Figure 18.3 The additional social (top) and environmental (bottom) layers of the triple-layer business model canvas from Joyce, Paquin and Pigneur (2015).

users. However, often the key beneficiaries (not the implementers) can also be the municipality as a whole (benefiting from NBS implementation), private businesses or schools, etc. (benefiting from NBS facilities), or from close proximity to implemented NBS.

- Key partners moved from the far left side (infrastructure) to the right half, next to key beneficiaries. This allows for a visualization of an overlap between key partners and key beneficiaries that occurs in some NBS.

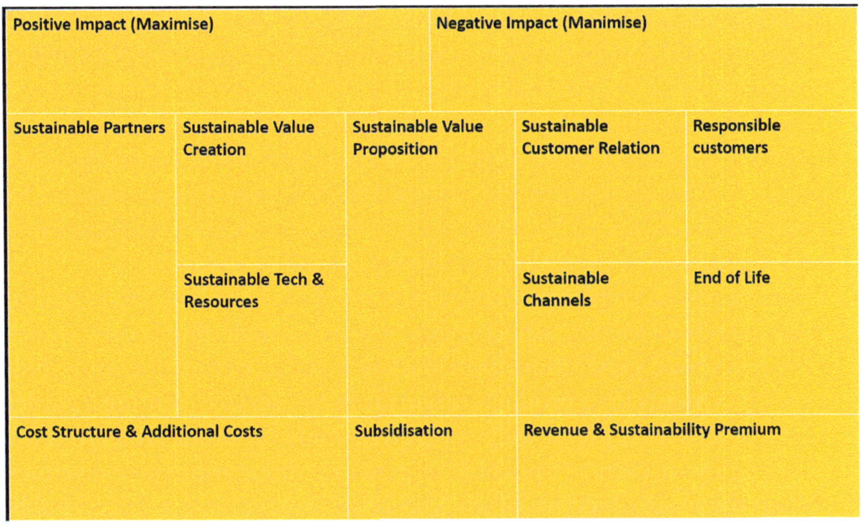

Business models in NBS **309**

Positive Impact (Maximise)			Negative Impact (Manimise)	
Sustainable Partners	Sustainable Value Creation	Sustainable Value Proposition	Sustainable Customer Relation	Responsible customers
	Sustainable Tech & Resources		Sustainable Channels	End of Life
Cost Structure & Additional Costs		Subsidisation	Revenue & Sustainability Premium	

Figure 18.4 Sustainable business model canvas from Gerlach (2015).

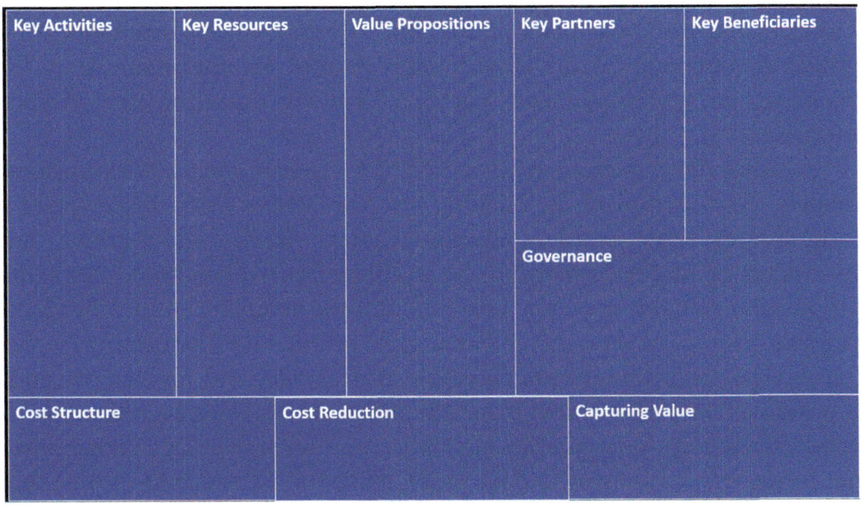

Key Activities	Key Resources	Value Propositions	Key Partners	Key Beneficiaries
			Governance	
Cost Structure		Cost Reduction	Capturing Value	

Figure 18.5 NBS business model proposed by the project Connecting Nature.

- Customer relationships and channels positioned between value propositions and customer segments in the traditional BMC are excluded here. Since NBS are usually implemented as green infrastructure, opposed to delivery of products and goods in traditional businesses, the delivery channels and customer relationship aspects receive less emphasis by dropping these blocks.

310   Nature-Based Solutions for Urban Renewal in Post-Industrial Cities

- Governance is added as a new building block into the adjusted BMC. This reflects the importance of identifying early on how the NBS will be managed on an operational basis.
- Cost reduction is added centrally in the bottom part next to cost structure (left) and capturing value (right). Cost reduction is key, especially when public entities implement and maintain NBS. Lower direct costs can, for example, be achieved by lower maintenance workload, self-sustaining and developing permaculture principles, or use of volunteers.
- Capturing value replaces revenue streams. This broadens the capturing from merely monetary aspects to value aspects, including financial ones. Direct revenue generation from NBS is a viable option for some NBS types (payments for a product [urban farming], service [renting a garden plot], or fees for a green wall/roof, etc.), while for other NBS types it is challenging or even not possible to generate direct monetary revenues. When the main objective of an NBS implementation is a shared public good, e.g., improved environmental benefits, financial revenues are in most cases only possible from public sources (funding incentives).

The adjusted BMC has been piloted with Connecting Nature cities. The project Connecting Nature disseminates its tool for communication, for planning, for identifying new partners, and for exploring new finance sources.

ProGIreg developed an own template for summarizing NBS business models. This template is building on the BMCs introduced before. It integrates some building blocks of other BMC modifications, but re-structuring and re-organizing the template (see Figure 18.6). The value proposition remains centrally positioned in the template. It includes both tangible (goods and services) and intangible values. Like Connecting Nature (see previous), governance is playing a crucial role, and for this reason it is positioned centrally below the value proposition. The governance block bundles together the organizational structure, ownership (e.g., cooperative, not-for-profit, privately owned for profit, and publicly traded for-profit), and decision-making policies of the NBS. Hierarchy, transparency, consultation, profit sharing, and other issues are subsumed under the building block of governance. To the right side, the customers and beneficiaries along with the relationships and channels are positioned. Customers are people, groups, or entities, who are paying for the values offered (value proposition), while the beneficiaries are not paying for the NBS values monetarily. Accordingly, the bottom part differentiates between two main ways of generating money for NBS maintenance and evolution, as well as paying back investment costs of the NBS. These are revenue streams and financing. While revenue streams

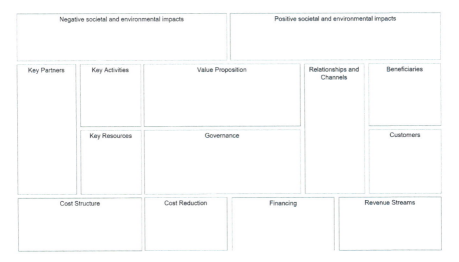

Figure 18.6 ProGIreg's BMC template.

cover money coming from customers, financing is required when the NBS is offering values for beneficiaries without getting paid for it. Financing is mainly realized by public funding. The left side keeps the structure of the traditional BMC. The key resources, activities, and partners are listed in the three building blocks, while the bottom left part focuses on costs (cost structure) and ways to reduce costs (cost reduction). The latter can be realized by reduced maintenance costs of the NBS (compared to other uses), use of volunteers during NBS implementation and/or maintenance. The top segment of proGIreg's BMC highlights the main positive (right) and possibly also negative (left) social and environmental impacts. This is coming from the sustainable BMC proposed by Gerlach (2015) (see above).

State-of-play on NBS business models

Since the late 2010s, EU research and innovation projects have contributed to the growing knowledge on NBS business models (Mayor et al., 2021). Accelerated by the COVID-19 pandemic, NBS implementations are challenged by budget allocation dilemmas for local governments. Although more and more project implementations are able to show NBS's multi-functionality and cross-sectoral benefits, 'public budgets for NBS investment are often insufficient to drive their mainstreaming' (Mayor et al., 2021: 2). Several studies (i.e., Bockarjova et al., 2020; Jacobs et al., 2016; Croci et al., 2021) demonstrate the value of nature and NBS in monetary and non-monetary terms. 'However, there remains a significant gap between articulating the value of nature and finding stakeholders who are actually willing to pay for nature' (Mayor et al., 2021: 2). Another crucial aspect is the predominant concentration on the capital investment phase only. Mayor et al. (2021) also argue that 'NBS appreciate over time but requiring ongoing financing of operational or stewardship costs' (Mayor et al., 2021: 2). This is different to grey infrastructure, which are depreciating with time. Thus, the generation of financial revenues is key for the sustainability, evolution, and expansion of NBS.

The EU project Naturvation developed a Business Model Catalogue for urban NBS, in which they present eight different business models by using their project specific approach (Toxopeus, 2019).

> Taking action on nature-based solutions does not only depend on establishing the right policy conditions and financial resources, but also on establishing business models that can ensure their sustainability over time. Nature-based solutions often create a complex array of public and private benefits, and developing business models that are able to capture and realise this value can be challenging.
>
> (Naturvation, 2019)

In their approach, they focus on four main building blocks of the business model, namely (a) value proposition, (b) value delivery, (c) value capture, and (d) enabling conditions and risks. Based on the Naturvation experiences, eight NBS business models are proposed in the catalogue:

- Risk reduction
- Green densification
- Local stewardship
- Green health
- Urban offsetting
- Vacant space
- Education
- Green heritage (Toxopeus, 2019).

Nature-based enterprises

A few years ago, the term nature-based enterprises was created to bring together the nature-based perspective of interventions and the business dimension of these activities.

> Nature-based enterprises (NBEs) use nature as a core element of their product/service offering. Nature may be used directly by growing, harnessing, harvesting or restoring natural resources in a sustainable way and/or indirectly by contributing to the planning, delivery or stewardship of sustainable nature-based solutions.
>
> (McQuaid et al., 2020: 5)

Due to their rootedness in nature, NBE embraces a huge variety of types. NBE include among others:

- Ecosystem creation, restoration, and management;
- Green buildings (e.g., living green walls and roofs);
- Public and urban spaces (e.g., urban forestry, gardens, etc.);
- Water management and treatment (e.g., wastewater management);
- Sustainable agriculture, food production (e.g., agroforestry, beekeeping, regenerative farm);
- Sustainable forestry and biomaterials; and
- Sustainable tourism, health, and well-being (e.g., agri- and eco-tourism, nature-based tourism).

NBE can also use nature indirectly. For instance, financial services offering carbon offsetting, natural capital accounting, and investment for biodiversity and conservation. Indirect use of nature can also be exploited in smart technology, monitoring and assessment; education, research, and innovation activities; advisory services (McQuaid et al., 2020).

Social enterprise

ProGIreg activities perform a quadruple helix approach bringing together government, academia, industry, and community. From an entrepreneurial perspective, this quadruple helix approach positions many NBS developments between traditional business and traditional charity measures. The need to define these organizations has also been recognized by policymakers; in 2017, the EU Commission defined social enterprises as organizations that combine social objectives with an entrepreneurial spirit and focus on achieving broader social, environmental, or community objectives (EU Commission, 2017). Defourny and Nyssens (2012) synthesized trends and developments of social enterprises from a European perspective, EMES – the European research network on social enterprises, including the overlapping and complementary working fields social entrepreneurship, social economy, social innovation, and solidarity economy. Their European view on social enterprises provides a set of indicators for three distinct dimensions:

- economic and entrepreneurial dimension;
- social dimension; and
- participatory governance.

The economic and entrepreneurial dimension consists of the indicators (a) continuous activity producing goods and/or selling services, (b) significant level of economic risk, and (c) minimum amount of paid work required.

For fulfilling the social dimension's criteria, (a) the explicit aim has to benefit the community, (b) the activity was launched with or by a group of citizens or civil society organizations, and (c) limited profit distribution, avoiding a profit maximization behaviour.

A high degree of autonomy, a decision-making power, which is not based on capital ownership and a participatory nature (involving various parties affected by the activity), defines the participatory governance dimension.

These three dimensions define social enterprises, which as of recently can be found in several branches, including the agriculture and food branch (Defourny and Nyssens, 2012; Martens et al., 2022).

The previously mentioned dimensions and characteristics of social enterprises show many overlaps with the quadruple helix approach and actions to establish NBS in proGIreg, namely, co-design, co-implementation, and co-maintenance. NBS can be developed not only from traditional businesses including profit maximization, but also from public entities solely relying on subsidies, funding, or grants. However, many NBS of proGIreg are positioned between these two. They incorporate all or some aspects of social enterprises defined by Defourny and Nyssens (2012). Social enterprises are positioned centrally in the triangle between market, state, and community (see Figure 18.7). As visualized in the figure, the individual setting differs between social enterprises; while some are closer to market/business, others are closer to civil society/community or public entities/state.

Martens et al. (2022) make use of the social enterprise approach to classify new hybrid cooperation models for short food supply chains in the urban-rural nexus. They follow the ARA concept (actor, resource, action) in their explorative study by positioning five community-based urban farming pilots into the triangle presented in Figure 18.7 including a forecast (see Figure 18.8).

Business models in NBS 315

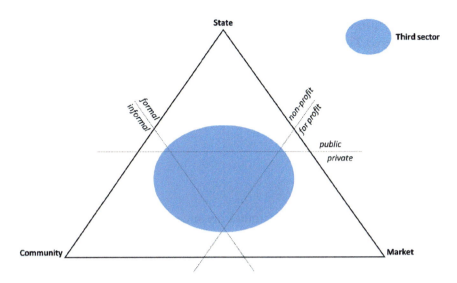

Figure 18.7 Social enterprises, the third sector, positioned between market, state, and community (adapted from Martens et al., 2022).

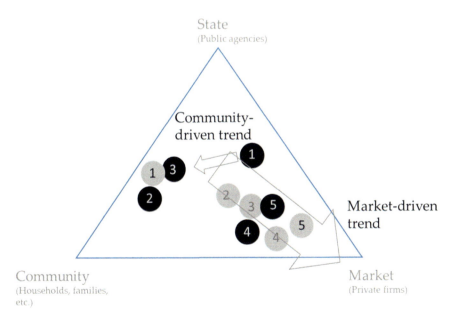

Figure 18.8 Five hybrid cooperation models in urban/peri-urban food production positioned between market, state, and community. Current position in black; forecast position (5 years) in grey (Martens et al., 2022).

316 Nature-Based Solutions for Urban Renewal in Post-Industrial Cities

For the future, the pilots exhibit a market-driven as well as a community-driven trend. This study shows the dynamics of social enterprises, which start somewhere between state, market, and community. Depending on the stakeholders involved, the objectives, and framework conditions (political, ownership, involvement, finances, etc.), the positions can change, either deliberately or otherwise.

Getting inspired: proGIreg case studies of NBS business models

The following proGIreg case studies are meant to illustrate the diversity of NBS business models without aiming for completeness or equilibrium. The examples from proGIreg's Front-Runner Cities are intended to serve as inspiration and consideration of business ideas when co-designing, co-implementing, and co-maintaining NBS.

The financial revenues for NBS can be manifold. While some NBS do not generate any direct monetary revenues, others are able to generate one-off purchases or recurring revenue stream models. From a municipal perspective, NBS have the potential to reduce maintenance costs.

One-off purchase

In the Torino's Living Lab, an aquaponics system was developed within an older greenhouse construction. Following an open call of the City of Torino, Vittorio Agù built the aquaponics system. His business is called "Mitte Garten" (Torino City Lab, 2022).

The system brings together fish production and vegetable production – here even in one floating tank system (see Figure 18.9), while most other systems separate fish and plants, meaning that they are grown and raised in separated tanks, while the water and nutrients are circulating between both components. In this setup, the fish eat the small plant roots, which reduces the disease pressure since most pathogens are entering through the root system. Aquaponics can operate on a relatively small space and for this reason can be integrated into inner-city boroughs.

Plants such as lettuce, basil, tomatoes, edible flowers, and zucchini float on top of the water. Water control is the most important part of the system, and specific pH and electrical conductivity controls are needed to be able to monitor plant and fish life and quality.

Business models in NBS 317

Figure 18.9 *"Mitte Garten"* aquaponics system in Mirafiori Sud, Torino. A video is available at https://progireg.eu/nature-based-solutions/aquaponics/.

All the products are sold locally in the Living Lab area as well as other districts of Torino. The main customers are restaurants adding aquaponics products into their offer and local population. Especially restaurants are willing to pay premium prices for these vegetables and fish.

Apart from the one-off purchase, Mitte Garten integrates local community, namely, schools and rehabilitation centres, through site visits.

Bundling all aspects of "Mitte Garten" together, the traditional one-off purchase of products – here vegetables and fish – is strongly interwoven with technological innovation and social cohesion and community building.

Recurring revenues

In Dortmund, another, much larger, aquaponics system is currently being under construction (see Figure 18.10). It consists of two greenhouses with two aquaponics systems under construction (state as of August 2022). While some concepts are comparable to the "Mitte Garten" NBS in Torino (research on technological innovation and optimization; Technology Readiness Level increase), the Dortmund aquaponics systems are looking at a completely different revenue stream and business model.

Figure 18.10 Situation of the aquaponics system Coking Plant Hansa Dortmund in August 2022 (Bernd Pölling).

One-off purchase business models rely on single purchases. By contrast, recurring revenue streams allow a higher level of income security when being successful in generating a sufficient amount of regular (weekly, monthly, yearly, etc.) payments. Dortmund's aquaponics system aims to copy and adapt the "rent-a-field" concept, in which urban and peri-urban farmers, situated in geographical proximity to residential areas, rent out small parcels for one season (Timpe, 2017). Notably in German-speaking countries, this concept has gained popularity among farmers and leasing gardeners. Long waiting lists for several "rent-a-field" examples confirm the high demand for food production concepts based on leasing concepts. The rental fees increase financial security for the farmers independently of the yield and product market prices. On the one hand, this provides security financially, but on the other hand, might reduce income in times of high demand and high prices. In proGIreg, the Dortmund aquaponics partner opted for the recurring revenues model. Small rafts with a variety of vegetables and herbs will be rented to interested city dwellers who are in return paying renting fees for the rafts. This copied version of the "rent-a-field" concept is the first of its kind with aquaponics' rafts. In case of little interest, part of the produce can also be sold to city dwellers or nearby restaurants or canteens. Another revenue stream being considered is sponsorship, by which companies will donate the production. In return, the food is sent to local food banks.

This recurring revenue stream means a shift from a producer and seller of food and other goods to a service provider. As service provider, the production itself loses importance, but creating an attractive location, good offers, and positive values is becoming part of core activities. In the aquaponics example, this means events and dissemination activities on vegetable and fish production, including barbecues, guided tours, and workshops. Thus, food production and selling is no longer a key activity.

This leasing concept is also practiced in Mirafiori Sud, Torino, in Cascina Piemonte. As described in the proGIreg project, Orti Generali was born with the aim of building 'an enterprise model for the transformation and management of post-industrial and metropolitan residual agricultural areas based upon ecological sustainability and social equity' (Saraco, 2021). Apart from renting gardens, Orti Generali pursues different objectives including education, social inclusion, and employment training (https://progireg. eu/turin/). The whole area including the gardens covers more than one hectare (>12,000 m²). In total, 160 gardens were created and more than 300 trees planted. The gardens are rented to private citizens, families, and collectives. The rental fee is lower for people under 35 years of age and for economically disadvantaged people. It took around four months to rent out all 160 garden plots. The rental fees differ between "standard gardens", the gardens dedicated for people under 35 years of age and the so-called social gardens for economically disadvantaged people. Depending on the size, the gardeners have to pay a monthly rent (see Table 18.1).

Additionally, the public space is no longer maintained by the city of Torino, but by the Association in charge. This measure reduces the maintenance costs for the municipality. So far, the recurring revenue model turned out to be successful considering a long waiting list of more than 150 entries.

Table 18.1 Monthly rents at Orti Generali

Rent (€ per month)	50 m²	75 m²	100 m²
Standard rent	25	35	45
Young rent (people < 35 years)	15	20	30
Social gardens	5		10

However, the business perspective is only one view on this specific topic, but also relates to the other NBS implementations. In all interventions, the financial issues (in many cases primarily) have to be considered in light of other social and environmental goals.

Cost reduction

The costs of maintaining public green space is a major cost item for municipalities and other public entities (Tempesta, 2015). In direct monetary value, it is a major cost item, but in many cases 'the flow of benefits largely exceeds the management costs, so urban parks seem to produce a net gain for the citizens' (Tempesta, 2015). These benefits, which are either non-monetary, or are not flowing directly back into the organization maintaining the parks, do not offer a direct financial return from an accounting perspective. However, as an important task, public money is used and valued for creating public goods. That being said, it has to be acknowledged that municipalities and public authorities are struggling with shrinking budgets. Thus, new ways of maintaining and developing further public spaces are urgently required. One way is co-designed, co-implemented, and co-maintained spaces as in proGIreg's Front-Runner Cities Dortmund and Turin. In Dortmund's Living Lab, green public spaces are experiencing a shift from lawns to biodiversity-rich flower meadows (see Figure 18.11).

Figure 18.11 Flower meadows in Dortmund's Living Lab (Rolf Morgenstern).

Green roofs are another NBS, which are able to contribute to more robust flood management. Water infiltration properties on green roofs help to retain water and lower food peaks in inner city, densely built-up areas, which are otherwise lacking water retention areas. In the Living Labs of Turin and Zagreb, green roofs have been created. This reduces flood damage costs and prevents cities from building costly installations, such as underground basins meant to retain water in case of heavy rainfall. It shows that cost reduction NBS models are not only possible for maintenance, but also during the implementation, including investment phases.

Lessons learnt

Technological barriers were usually overcome within manageable time frames, when a dedicated group of stakeholders and knowledgeable experts was established to actively search for solutions. The challenge from this perspective is how to create strong project groups and to facilitate and support their collaboration, co-design, and co-implementation. It required considerable project effort to promote participation in implementing NBS, and to encourage groups of interested citizens to undertake tasks and adopt projects as their own.

Formalizing and strengthening the typically informal, loose, and organic network of citizens can be a sensible path in supporting this goal. For instance, during the project period, the Dortmund Food Council was established. While this movement was not created or assisted by the proGIreg project itself, it became apparent during the second half of the project that the networking benefits provided by city food councils functioned well in this regard.

Stakeholders involved in the co-design process tended to focus on the NBS implementation, yet, neglected the following operational and maintenance phase of the implemented solution. It is crucial to also consider and develop the economic perspective right from the beginning. Therefore, technical or administrative issues should not be prioritized over financial ones. While implementation might be achieved without the development of a convincing and locally adjusted business model for NBS implementation, the ongoing maintenance might not be successful.

In this regard the mantra of "keeping it simple" really pays off. Simple measures like flower meadows were not only easy to implement, but their temporary nature also made it much easier to secure the required spaces. Solutions that feature infrastructure (e.g., requiring a building permit, perennial plants or trees and shrubs) require the space to be secured for longer periods. This might involve land leasing contracts and all the negotiations and financial issues that come with them. Solutions that require little maintenance after the implementation phase might not require a strong economic oriented business model.

For example, food forests do not necessarily need maintenance after they are established. Experience from older food forests outside of the proGIreg project shows that even when neglected for several years, the plant structure and functions prevail. Since permaculture involves design of systems with the explicit goal of low maintenance, this design method can and should be an inspiration when searching for sustainable NBS.

Business models for NBS differ from conventional businesses since instead of having one clearly defined customer, NBS usually have several different beneficiaries. NBS benefit society in general, making it hard to define the specific beneficiaries of a solution. When scouting for a sensible business model for an NBS, it therefore makes sense to consider how secondary beneficiaries could be integrated into financing the implementation and operation of the NBS.

One of the major barriers identified in the proGIreg project was the administrative processes and regulations, as well as the administrative body itself. Regulations regarding a wide range of issues, from parking spaces to rainwater management, are often formulated in a language that leaves very little leeway for the municipal employee in charge. In addition, NBS are usually not explicitly mentioned and managed within the legal framework. There are simply no categories for NBS. Administrative procedures exist for grey infrastructure, but few if any for (productive) green infrastructure. As a consequence, the municipal employees have to fit the NBS into the existing framework, based on requirements that are not actually suitable in terms of the solution.

For example, the aquaponics greenhouse system in Dortmund had to be treated like any other larger scale building, prompting the requirement for several parking places. This is in stark contrast to the NBS concept which is aiming to recruit citizens from the direct vicinity of the location, within a walkable distance. Only following a lengthy solution-seeking process could the administration be persuaded to treat this aspect of the building in a similar way to an allotment garden, thereby largely reducing the requirement for parking spaces.

Incorporating NBS into official city policies and integrating them into the city regulations framework (while additionally wording the regulations in a way that leaves room for interpretation by municipal framework users) is considered an important step towards a wider adoption of NBS in cities. There is a lack of insight that *not doing anything wrong* does not mean that *the right thing* is being done, leaving a plethora of potentially good solutions completely off the table.

A Food Policy Council can act as a conduit to solve the issues described. The urban citizens' side of such a council can increase information flow and nurture an interested and engaged society. With regards to the administration, established communication partners vastly simplify and streamline the otherwise vague task of "involving the public". Founding a Food Policy Council to tackle the growing requirement of involving citizens early in city planning processes is a recommended first step.

References

Bockarjova, M., Botzen, W.J.W. and Koetse, M.J. (2020) Economic Valuation of Green and Blue Nature in Cities: A Meta-Analysis. *Ecological Economics*, 169, 106480.

Boons, F. and Lüdeke-Freund, F. (2013) Business Models for Sustainable Innovation: State-of-the-Art and Steps towards a Research Agenda. *Journal of Cleaner Production*, 45, 9–19.

Connecting Nature (2019) Nature-Based Solutions Business Model Canvas Guidebook. https://connectingnature.eu/sites/default/files/downloads/NBC-BMC-Booklet-Final-%28for-circulation%29.pdf.

Croci, E., Lucchitta, B. and Penati, T. (2021) Valuing Ecosystem Services at the Urban Level: A Critical Review. *Sustainability*, 13, 1129.

Defourny, J. and Nyssens, M. (2012) The Emes Approach of Social Enterprise in a Comparative Perspective. EMES network WP no. 12/03. https://www.emes.net/content/uploads/publications/EMES-WP-12-03_Defourny-Nyssens.pdf.

EU Commission (2017) Social Enterprises. https://single-market-economy.ec.europa.eu/sectors/proximity-and-social-economy/social-economy-eu/social-enterprises_en.

Faivre, N., Fritz, M., Freitas, T., de Boissezon, B. and Vandewoestijne, S. (2017) Nature-Based Solutions in the EU: Innovating with Nature to Address Social, Economic and Environmental Challenges. *Environmental Research*, 159, 509–518.

Ferranti, E. and Jaluzot, A. (2020) Using the Business Model Canvas to Increase the Impact of Green Infrastructure Valuation Tools. *Urban Forestry & Urban Greening*, 54, 126776.

George, G. and Bock, A.J. (2011) The Business Model in Practice and Its Implications for Entrepreneurship Research. *Entrepreneurship Theory and Practice*, 35(1), 83–111.

Gerlach, R. (2015) The Sustainable Business Model Canvas. Threebility: Tools for Sustainable Innovation. https://www.threebility.com/post/the-sustainable-business-model-canvas-a-common-language-for-sustainable-innovation.

Henriksen, K., Bjerre, M., Almasi, A.M. and Damgaard-Grann, E. (2012) Green Business Model Innovation. Conceptualization Report. Nordic Innovation Publication. https://www.nordicinnovation.org/Global/_Publications/Reports/2012/2012_16%20Green%20Business%20Model%20Innovation_Conceptualization%20report_web.pdf.

Jacobs, S., Dendoncker, N., Martín-López, B., Barton, D.N., et al. (2016) A New Valuation School: Integrating Diverse Values of Nature in Resource and Land Use Decisions. *Ecosystem Services*, 22, 213–220.

Johnson, M.W., Christensen, C.M. and Kagermann, H. (1996) Reinventing Your Business Model. *Harvard Business Review*, 9(10), 57–68.

Joyce, A. and Paquin, R.L. (2016) The Triple Layered Business Model Canvas: A Tool to Design More Sustainable Business Models. *Journal of Cleaner Production*, 135, 1474–1486.

Martens, K., Rogga, S., Zscheischler, J. Pölling, B., Obersteg, A. and Piorr, A. (2022) Classifying New Hybrid Cooperation Models for Short Food-Supply Chains—Providing a Concept for Assessing Sustainability Transformation in the Urban-Rural Nexus. *Land*, 11, 582. https://doi.org/10.3390/land11040582.

Mayor, B., Toxopeus, H., McQuaid, S., Croci, E. et al. (2021) State of the Art and Latest Advances in Exploring Business Models for Nature-Based Solutions. *Sustainability*, 13(13), 7413.

McQuaid, S., Kooijman, E. and Fletcher, I. (2020) Nature-Based Enterprise Guidebook. Connecting Nature Project. https://connectingnature.eu/sites/default/files/images/inline/Enterprise.pdf.

Naturvation (2019) Creating Business Models. https://naturvation.eu/action/creating-business-models.html.

Osterwalder, A. (2004) The Business Model Ontology. A Proposition in a Design Science Approach. Dissertation Thesis. University of Lausanne, Switzerland.

Osterwalder, A. and Pigneur, Y. (2009) *Business Model Generation*. Strategyzer Series, Zurich, Switzerland.

Saraco, R. (2021) Implementation Monitoring Report No. 2. Del. 3.4. proGIreg. https://progireg.eu/resources/planning-implementing-nbs/.

Tempesta, T. (2015) Benefits and Costs of Urban Parks: A Review. https://doi.org/10.13128/AESTIMUM-17943.

Timpe, A. (2017) Produktive Parks entwerfen. Geschichte und aktuelle Praxis biologischer Produktion in europäischen Parks. Dissertation thesis. RWTH Aachen University.

Torino City Lab (2022) Mitte Garten. https://www.torinocitylab.com/en/mitte-garten.

Toxopeus, H. (2019) Taking Action for Urban Nature: Business Model Catalogue. Naturvation Project. https://naturvation.eu/sites/default/files/results/content/files/business_model_catalogue.pdf.

Index

Note: **Bold** page numbers refer to tables; *italic* page numbers refer to figures and page numbers followed by "n" denote endnotes.

Agence Ter studio 48
agro-ecosystem 41, 42
Agù, V. 316
alcohol-related habits 235
animal biodiversity 39–42
anxiety symptoms 236
aquaponics system 32–33, *32*; in
 Dortmund Living Lab 70–71, **72–73**,
 74, 75; Turin Living Lab 158–159;
 Zagreb Living Lab 85–89, 96–98
ARA concept (actor, resource, action)
 314

Barbera, F. 189
Barbier, E. 209
barriers, for NBS implementation 302;
 Dortmund Living Lab 22; types of
 303; Zagreb Living Lab **95**
BINs *see* Biodiversity Indicators for
 NBS (BINs)
biodegradability 28
BIODIVERSA+ 289
biodiversity: animal 39–42; assessment
 of 215–218; benefits of 215–227;
 conservation 36, 37, 39–41, 158,
 216–218, 273; definition of 36; design

strategies 42–52; enhancement
 219; exposure 232–233; loss 8, 24,
 37, 42, 192, 216, 232, 288, 289,
 291–294; methodology of 218–221;
 nature-based solutions for 36–54;
 plant 42–52; pollinator 14–17, *15*,
 136, 143, 209, 212, 213, 218–220
Biodiversity Indicators for NBS (BINs)
 217
Biodiversity Strategy: 2011–2020 289;
 2030 289
biotopes, construction of 48–52
BMC *see* business model canvas (BMC)
business model canvas (BMC): building
 blocks of 306; Connecting Nature
 307–310, *309*; structure of *306*;
 sustainable *309*; template *311*; triple
 layered business model 307, *308*
"Business Model Generation" 306
business models 301–323; cost
 reduction 320–321; one-off purchase
 316–317; proGIreg case studies
 of 316–321; recurring revenues
 317–320, *318, 319*; state-of-play on
 311–312; thinking 304–305; tools
 305–311

326 Index

CALs *see* CLEVER Action Labs (CALs)
capitalism 24
carbon offsetting 287
Cascais: basic information *257*; challenges to 258–260; co-creation process 260–262, *262*; Final Urban Plan 263, *263*; Marianas River 261, *261*, 262; story of 257–258; URA conditions *258, 259*, 260
Catarino, R. 41
CBD *see* Convention of Biological Diversity (CBD)
circular economy: benefits of 207–214; biodegradability 28; costs 209–211; EU taxonomy of 29–30; leisure activities and clean energy on former landfills 30–33, *32*; methodology of 208–212; natural capital 28, 29, 31; nature-based solutions for 24–34; re-usability 28; waste valorization 26–27
clean energy on former landfills 30–33, *32*
CLEVER Action Labs (CALs) 145, 146
CLEVER Cities project 130, 134; co-creation of NBS for social inclusion in 145–146
climate adaptation 38, 39
climate change 1, 2, 5, 8, 24, 28, 30, 36, 37, 41, 289–293; adaptation 29, 286, 292; advancing 25; impacts 29; mitigation 3, 8, 31, 38–39, 42, 287, 294
Cluj-Napoca: basic information *264*; challenges to 265; co-creation process 269–270; Final Urban Plan 270–271, *271*; story of 264–265; URA conditions 265–269, *266–268*
CNEP *see* Connecting Nature Enterprise Platform (CNEP)
co-creation 122, 124, 249; barriers to 136–137, 139–140, *139*; Cascais 260–262, *262*; Cluj-Napoca 269–270; community-based urban farms and gardens *14*; definition of 125; demarcating, from stakeholder engagement 124–126; Dortmund Living Lab *21*; for Edible City Solutions 146–147; governance

models for 130, *130*, **131–132**, 141–142; phases of 129, *129*; Piraeus 276–277; for social inclusion in CLEVER Cities project 145–146; stakeholder engagement, intensity of 135–136, *137*; stakeholder engagement *versus* 125; target stakeholder groups, identifying and interacting with 132–135, *133*; Zenica 283
co-design 129, 252; NBS, with post-industrial communities 121–149, *122*; process-oriented approach of 20, *21*; workshops 251, *251*; Zagreb Living Lab 89–93
co-governance **131**
co-implementation 129, 140, 252
Coldiretti (Federation of Italian Farmers) 142
Collaboration Pact 141
co-maintenance 129
co-management 129, **131**; of Zagreb Living Lab therapeutic garden 143–144
community-based urban farms and gardens: barriers 20–22; biodiversity projects 17–19; challenges 20–22; co-creation cycle *14*; knowledge transfer 75; opportunities 20–22; participation events *19*; on post-industrial sites 82, 84–85; social innovation 10–14, *11–13*
community-driven trend *315*, 316
Connecting Nature: BMC modification 307–310, *309*
Connecting Nature Enterprise Platform (CNEP) 293
Continuing Professional Development (CPD) 293
Convention of Biological Diversity (CBD) 294; COP12 (2014) 286; COP14 (2018) 286, 290; COP15 (2022) 288
co-occurrence network analysis *39*
Cordis 248
costs reduction 320–321
COVID-19 pandemic 12, 17, 22, 58, 91, 93, 97, 142, 145, 182, 193, 294, 295, 311; impact on Dortmund Living

Lab 20, 70; impact on Turin Living Lab 162–163
CPD *see* Continuing Professional Development (CPD)

decarbonization 37, 88, 289, 290
"Decent Work in Nature-based Solutions" 288–289
Defourny, J. 313, 314
Deksissa, T. 37–38
depression 236
Design Capital 152
"die Urbanisten e.V." 11, 16, 18
Dortmund Food Council 321
Dortmund Living Lab 57–76, *59*, *61*; administrative barriers 22; aquaponics system 70–71, **72–73**, *74*, *75*, 317–318, *319*; characteristics of **59**; circle economy 30–31; co-creation cycle *21*; co-design process 20, 21; cost reduction 320, *320*; COVID-19 pandemic, impact of 20, 70; knowledge transfer 71, 74–76; map 66, *67*; NBS implementation 63, *63*, **64**; NBS implementation, challenges and opportunities for 66–70, **68**; NBS implementation into existing policy framework, integration of 65; participation events **19**; policy frameworks, impact of 60, 62–63, *62*; pollinator diversity with local communities, leveraging 142–143; social innovation 9, *10*, 15, 16; soil contamination 66, 68, 69; space scarcity 20; stakeholder collaboration 65–66; success of 22–23; as testing ground for NBS 57–58; trust building 21
doughnut economy 24, *26*
Durban: Community Ecosystem Based Adaptation (CEBA) 286

eBMS *see* European Butterfly Monitoring Scheme (eBMS)
EcoJardin label 46, 50, 54n2
ecological niches, construction of 48–52

ecological resilience 37, 38, 161
Economic and Labour Market Questionnaire (ELMQ) 187, 208
ECS *see* Edible City Solutions (ECS)
edge effects 104
Edible Cities Network 146
Edible City Solutions (ECS): co-creation of 146–147
EdiCitNet 130, 146–147
eDNA *see* environmental DNA (eDNA) sequencing
Eichler, G. M. 190
Ellen MacArthur Foundation 27
ELMQ *see* Economic and Labour Market Questionnaire (ELMQ)
EMES 313
ENACT Partnership for Nature-based Solutions 288
energy / fatigue 235–236
environmental DNA (eDNA) sequencing 221
EU *see* European Union (EU)
EU Commission 313
EU-PoMS *see* European Pollinator Monitoring Scheme (EU-PoMS)
EuroCities 292
EuroCities 292
European Butterfly Monitoring Scheme (eBMS): Citizen Science project 222
'European Capital of Innovation' competition (2016) 152
European Circular Economy Action Plan 29
European Commission 7, 8, 217–218, 292; on biodiversity 40; on circular economy 28–29; nature-based solutions, definition of 2, 207, 305; NBS Action Plan 217, 218; social innovation, definition of 189
European Pollinator Monitoring Scheme (EU-PoMS) 219
European Union (EU) 2, 185; Adaptation Strategy 289; Biodiversity Strategy for 2030 289; on circle economy 29–30; Green Deal 31, 82, 289; Research Programme Horizon 2020 (2014–2020) 5, 134, 148, 180, 289

328 Index

Farfalle in Tour project 224–225
FBNC *see* Forestry Bureau of Ningbo
  City (FBNC)
FC *see* Follower Cities (FC)
Ferranti, E. 307
Ferreira, V. 192
first nature 43
Follower Cities (FC): Cascais *see* Cascais;
  Cluj-Napoca *see* Cluj-Napoca; Final
  Urban Plan 253–254; outcomes
  116, **117–118**; Piraeus *see* Piraeus;
  replication strategies for greening
  urban environments 247–284;
  "Roadmap toward Urban Planning
  for NBS" 249–253, *250–253*; spatial
  analysis of 103; Zenica *see* Zenica
Food Forest, Dortmund 136
Food Policy Council 323
Forestry Bureau of Ningbo City
  (FBNC) 173
fossil energy 1
Frantzeskaki, N. 123
FRC *see* Front-Runner Cities (FRC)
Front-Runner Cities (FRC) 145, 152,
  177, *255*, 302, 316; health and
  wellbeing 238, 241; outcomes 115,
  115; spatial analysis of 103

GAD-7 *see* Generalized Anxiety
  Disorder questionnaire (GAD-7)
Galego, D. 190
GDS-5 *see* Geriatric Depression Scale
  (GDS-5)
Generalized Anxiety Disorder
  questionnaire (GAD-7) 236
General Questionnaire (GQ) 234, 241
Geriatric Depression Scale (GDS-5) 236
Gerlach, R. 307, *309*, 311
Glasgow Climate Pact 287
governance models: for co-creation
  130, *130*, **131–132**, 141–142
government actor–led model *138*
GQ *see* General Questionnaire (GQ)
Green and Blue Sesvete (ZIPS) 79–81,
  82, 89, 91, 93, 144
Green Corridors: Turin Living Lab
  159–160; 222–224, *225*, *226*; Zagreb
  Living Lab 81–82, *82*

Green Deal 31, 82, 289
green infrastructures 9, *38*, 144, 147,
  155, 161, 166, 167, 207, 209, 303, 307;
  Cluj-Napoca 264–271, *264*, *266–268*,
  271; Dortmund Living Lab 57;
  governance models of 130; Piazzale
  Aldo Moro, Turin 53; spatial analysis
  of 106; Zagreb Living Lab 87–89,
  91, 93
green roofs: Turin Living Lab 159, *160*,
  195; Zagreb Living Lab 85–89, *86*,
  96–98, *97*
green walls: Piazzale Aldo Moro, Turin
  51–52, *51–53*; Turin Living Lab 159,
  160, *195*; Zagreb Living Lab 85–89,
  86, 96–98, *97*
Grün Berlin 44

health: assessment of 230–234; benefits
  of 230–241, *231*; case studies 238–241;
  methodologies of 234–238
Health Impact Assessment (HIA) 185, 241
Healthy Green Recovery from
  COVID-19 294
Heritage Management Department,
  Turin 142
Horizon Europe (2021–2027) 289;
  UP2030 98
Horlings, L. G. 192

ICLEI Europe 292
ILO *see* International Labour
  Organization (ILO)
Infrastructure for Spatial Information
  in Europe (INSPIRE) Directive
  (2007) Data Themes 109
insect pollinators, biodiversity
  indicators for 216
INSPIRE *see* Infrastructure for Spatial
  Information in Europe (INSPIRE)
  Directive (2007) Data Themes
Intergovernmental Panel on Climate
  Change Report (IPCC) 8
International Garden Exhibition Ruhr
  2027 60, 62
International Labour Organization
  (ILO): "Decent Work in
  Nature-based Solutions" 288–289

Index **329**

International Physical Activity Questionnaire (IPAQ) 236
International Union for Conservation of Nature (IUCN) 40, 215; "Decent Work in Nature-based Solutions" 288–289; Global Standard for Nature-Based Solutions 287
IPAQ *see* International Physical Activity Questionnaire (IPAQ)
IPCC *see* Intergovernmental Panel on Climate Change Report (IPCC)
IUCN *see* International Union for Conservation of Nature (IUCN)

Jaluzot, A. 307
Jardin des oiseaux, Paris 45–46, *47*, *48*
Joyce, A. *308*

Kabisch, N. 192
Kirchherr, J. 208
Kowarik, I. *43*
Kunming Action Agenda 290

LCA *see* life-cycle assessment (LCA)
Leaders' Pledge for Nature (LPN) 290
Learning Landscapes 2021 93, *94*
Lehmann, S. 38
leisure activities on former landfills 30–33, *32*
LE:Notre Landscape Forum 2019 93
life-cycle assessment (LCA) 187
Living Labs (LL) 3–7, 302; co-designing NBS in 132–140; Dortmund *see* Dortmund Living Lab; government and non-government roles, spectrum of **126**; Ningbo *see* Ningbo Living Lab; transformation, planning 252; Turin *see* Turin Living Lab; Zagreb *see* Zagreb Living Lab
LPN *see* Leaders' Pledge for Nature (LPN)

Mackinnon, K. 36
major life events 236
Martens, K. 314
Mayor, B. 312
mental health 144, 167, 233; and wellbeing 235–236

Mirafiori Onlus 142
Mitte Garten 316, *317*
Mittermeier, R. A. 36
Momentum for Change Initiative (2017) 286
monitoring 95, 96, 110, 113, 119, 158, 161, 184, *184*, 185, *186*; biodiversity 218, 226–227, *227*; data 95; definition of 179; environmental 159; long-term 217; plan 180–183; plankton biodiversity 220–221; pollinator biodiversity 219–220; psychological 190; social 146, 190; soil pollution 168; soil quality 177; stations 164; tools 238, **239–240**; water quality 175

Nationally Determined Contributions (NDCs) 287
natural capital 2, 28, 29, 31; accounting 313; investment 295
natural water retention measures (NWRM) 289
nature-based agriculture 41
nature-based enterprises (NBEs) 293, 312–313
nature-based solutions (NBS): benefits of 6; for biodiversity *see* biodiversity; for circle economy 24–34; co-design, with post-industrial communities 121–149, *122*; definition of 2, 36, 40, 207, 215, 288, 305; effectiveness, challenges in assessing 217; evidence-based benefits 179–187, *184*, *186*; Living Labs for *see* Living Labs (LL); policymaking 285–296; for social innovation 8–23; spatial analysis processes 101–120; in Turin Living Lab 156–162, *157*, *158*, *160*, *161*; types of 4–5, *5*; to urban regeneration 1–3, 6, 58, 59, 63, 66, 71, 76, 79, 91; *see also individual entries*
Naturfelder Dortmund e.V. 17, *17*, 143, 213, 214
Naturfelder Issum e.V. 16, 22
Naturpark Schöneberger Südgelände, Berlin 43–45, *44–46*

330 Index

Naturvation 307; Business Model Catalogue for urban NBS 312
NBEs *see* nature-based enterprises (NBEs)
NBS *see* nature-based solutions (NBS)
NBS c the DUAL s.r.l. 157
NBS-Visitor Questionnaire 200–206, 234, **239–240**
NDCs *see* Nationally Determined Contributions (NDCs)
Network Nature 293
Ningbo Living Lab 165–178; aquatic plants, planting **170**, 172–174, *173*, *174*; biodiversity monitoring 226–227, *227*; environmental compensation, procedures for **170**, 174–177, *176*, **176**; knowledge transfer 177–178; lake sediment into soil fertilizer, transforming 170–172, **170**, **171**, *172*; location of *166, 167*; NBS implementation, challenges and goals of 169–170; policy context, impact of 168–169; as testing ground for NBS 165
non-government actor-led model *132*, 137
NWRM *see* natural water retention measures (NWRM)
Nyssens, M. 313, 314

Obama Administration 286
obesity *235*
one-off purchase 316–317
Opportunities to Accelerate Nature-based Solutions: A Roadmap for Climate Progress, Thriving Nature, Equity and Prosperity (2022) 286
Orti Alti 159
Orti Generali urban garden, Turin 136, 213, 222, *223*, *224*; monthly rentals at 319, **319**
Osterwalder, A. 304, 306, 307

Paquin, R. L. *308*
Parc de Billancourt, Boulogne-Billancourt 48–50, *49*, *50*
Parco del Nobile Association 142
Paris Agreement on Climate Change 287, 290
Paris Climate Agreement 8

Parra, C. 191
perceived stress 236
Perceived Stress Scale (PSS-4) *236*
physical activity levels 236
physical health 144, 167
phytoplankton: biodiversity indicators with 216–217; biodiversity monitoring of 220–221
Piazzale Aldo Moro, Turin: green walls 51–52, *51–53*
Pignatti, S. 220
Pigneur, Y. 306, 307, *308*
Piraeus: basic information *272*; challenges to 273–275; co-creation process 276–277; Final Urban Plan 277, *278*; story of 272–273; URA conditions *273–276, 275–276*
planetary boundaries 24–25, *25*
plant biodiversity 42–52
policymaking 285–296
pollinator biodiversity 14–17, *15*, 136, 143, 209, 212, 213, 218; monitoring 219–220; Turin Living Lab 161
pollinator diversity: with local communities, leveraging 142–143
post-industrial cities: co-design NBS with 121–149; community-based urban farms and gardens 82, 84–85; Turin Living Lab 153–154, *154; see also individual entries*
proGIreg 5, 11, 15, 16; assessment domains *187*; BMC template *311*; co-creation, definition of 125; Eastern European Follower cities 148; key assessment domains *104*; Living Labs *see* Living Labs (LL); nature-based urban renewal approach 57; NBS implementation in Dortmund Living Lab 63, *63*; research cities *3, 4*; social innovation 194–196; spatial analysis component, methodology of 103–120; *see also individual entries*
PSS-4 *see* Perceived Stress Scale (PSS-4)

quadruple helix approach 111, 118, 122, *123*, 126–130, *127*, 134, 156, 163, 313, 314

Index 331

Raworth, K. 24
Raymond, C. M. 208
recycling 27, 33, 208
regenerated soil 31, 157, 254
ReGreen Zagreb 2021 93
rent-a-field 318
replication strategies, for greening urban environments 247–284; challenges and learning 254–256, *255*; roadmap's usefulness, critical reflection of 256; *see also* Follower Cities (FC)
Research Programme Horizon 2020 (2014–2020) 5, 134, 148, 180, 289
resilient city 38
re-usability 28
"Roadmap toward Urban Planning for NBS" 249–253, *250–253*; co-design 252; co-implementation 252; LL transformation, planning 252; preparatory work 251–252; roadmap's usefulness, critical reflection of 256

St. Urbanus community center: community-based urban farms in 11–13, *11–13*
Sandy storm (2012) 286
Schumpeter, J. 189
Schwarz, E. J. 190
SCL-90r *see* Symptom Checklist-90-Revised version (SCL-90r)
SDGs *see* Sustainable Development Goals (SDGs)
Secco, L. 191
Secretariat of the Convention 36
Seddon, N. 37, 40–41
self-perceived health 235
Sendai Framework on Disaster Risk Reduction 2015–2030 286
SF-36 *see* Short Form of the Self-Reported Health Questionnaire (SF-36)
Shannon Diversity Index 218, 220
Shannon Evenness Index 218, 220
Sharm El Sheik Implementation Plan (the SHIP) 288, 290

SHIP, the *see* Sharm El Sheik Implementation Plan (the SHIP)
Short Form of the Self-Reported Health Questionnaire (SF-36) 235
smoking 235
social cohesion 71, 91, 146, 147, 153, 163, 181, 233, 294, 317
social enterprises 313–316, *315*; definition of 313
social innovation 8–23; assessment of 190–193; benefits of 189–196; definition of 189; indicators of 193–194; methodology of 193–194
social justice 63, 124, 181, 230
societal challenges (SCs) 179, 180
socio-ecological resilience 8
somatization 235
SOPARC *see* System for Observing Play and Recreation in Communities (SOPARC)
space scarcity 20
spatial analysis process 101–120; basic data 110; common framework, building 101–102; follower cities 103; front-runner cities 103; methodology of 105–107, *109*; outcomes 115–116, **115**, **117–118**; plan and policy framework 111; research questions **108**; spatial data indicators 112; stakeholder identification 111; structural components of 110–114; SWOT analysis 112–113, *114*
SPAZIOWOW 141, 142
Spijker, S. N. 191
stakeholder engagement: co-creation *versus* 125; demarcating co-creation from 124–126; intensity of 135–136, *137*
Stefanakis, A. I. 29, 208
Strategy for Adaptation to Climate Change in the Republic of Croatia 82
sustainable development 208
Sustainable Development Goals (SDGs) 8, 39, 190, 294; SDG 11 38
Sustainable Energy and Climate Action Plan of the City of Zagreb 82
Symptom Checklist-90-Revised version (SCL-90r) 235

332 Index

System for Observing Play and Recreation in Communities (SOPARC) 234, 236–238, *237–238*, **240**, 241

Technology Readiness Levels (TRL) 248
Tianhe Aquatic Ecosystem Engineering Co., Ltd. 174
Torino Living Lab 316
transformative change, through NBS implementation 291–296
triple layered business model 307, *308*
TRL *see* Technology Readiness Levels (TRL)
Turin Living Lab 152–164; aquaponics 158–159; challenges to 161–162; characteristics of **153**; cost reduction 321; COVID-19 pandemic, impact of 162–163; green corridors 159–160; green roofs and walls 159, *160*, *195*; nature-based solutions in 156–162; new soil 157–158; opportunities of 161–162; policy frameworks, impact of 155–156; pollinators biodiversity actions 161; in post-industrial Mirafiori Sud district 153–154, *154*; strategic public–private partnership 160–161; urban community gardens 158, *159*

UIP *see* Urban Innovation Partnership (UIP)
Ulug, C. 192
UN *see* United Nations (UN)
UN Climate Action Summit (2019) 286
UNDRR: "Words into Action Nature-based Solutions for Disaster Risk Reduction" 291
UNEA *see* United Nations Environment Assembly (UNEA) 5.2
UNEP *see* United Nations Environment Programme (UNEP)
UN FAO, on circle economy 32
UNFCCC *see* United Nations Framework Convention on Climate Change (UNFCCC)
UN General Assembly 286

United Nations (UN): Biodiversity Conference (COP 15) 34; Climate Action Summit 2019 290; "State of Finance for Nature" 30, *33*; World Water Development Report on Nature-based Solutions (2018) 290
United Nations Environment Assembly (UNEA) 5.2 287
United Nations Environment Programme (UNEP): "Decent Work in Nature-based Solutions" 288–289
United Nations Framework Convention on Climate Change (UNFCCC) 294; COP17 286; COP21 287; COP26 37, 287; COP27 288
Università degli Studi di Torino 51
University of Turin: Department of Agricultural, Forest and Food Sciences 157; Department of Chemistry 157
University of Zagreb 144
Urban Greening Plan Guidance and Toolkit 292
Urban Greening Platform 292
Urban Innovation Partnership (UIP) 145
urban regeneration 98–102, 105, 110–112, 116, 118, 119, 124, 130, 133, 134, 140, 144, 148, 153, 248, 249, 280; area 275–276; nature-based solutions to 1–3, 6, 58, 59, 63, 66, 71, 76, 79, 91; post-industrial 5, *5*
urban regeneration process 98–100

"Vesela motika" (Happy hoe) company 97
Voorberg, W. H. 125

Walkability Index 185
waste valorization 26–27
WEF *see* World Economic Forum (WEF)
wellbeing: assessment of 230–234; benefits of 230–241, *231*; case studies 238–241; methodologies of 234–238
We Mean Business Coalition 293
WHO *see* World Health Organization (WHO)

Wilson, E. O. 42
World Conference on Disaster Risk
  Reduction, Sendai (2012) 286
World Economic Forum (WEF) 290
World Health Organization (WHO)
  294

Yilianhuimo Information Technology
  Co., Ltd. 174

Zagreb Living Lab 78–100; access to
  running water 86–87; aquaponics
  system 85–89, 96–98; barriers to
  95; bicycle lanes 86–87; challenges
  of 95–96; City Office for Economy,
  Environmental Sustainability and
  Strategic Planning 85; co-design
  activities 89–93; community
  building through NBS 78–79; cost
  reduction 321; Green and Blue
  Sesvete Green Corridor plan 81–82,
  82; green corridors 86–87, *87*, *88*;

green roofs and walls 85–89, *86*,
  96–98, *97*; low-carbon guidelines
  into new strategic documents
  88–89; NBS implementation 82–89,
  83, **83**; opportunities of *96*; policy
  frameworks, impact of 80–82; road
  and rail infrastructures *81*; in Sesvete
  neighborhood 79–80, *80*; Sljeme
  meat factory 79, *80*; therapeutic
  garden 84–85, *85*, *90*, *92*, 136;
  therapeutic garden, co-management
  of 143–144; urban regeneration
  process 98–100
Zenica: basic information *279*;
  challenges to 280; co-creation
  process 283; Final Urban Plan
  283–284, *284*; Green City Action
  Plan 280; story of 279–280; URA
  conditions 280–283, *281*, *282*
Ziegler, R. 190
zooplankton, biodiversity indicators
  with 216–217